Circuit Analysis Exam File

Artice M. Davis, San Jose State University, Editor; Peter Aronhime, University of Louisville; M. Awipi, Tennessee State University; J. R. Carver, Queensborough Community College; C. H. Chen, Southeastern Massachusetts University; Thomas J. Dorsett, Gonzaga University; Rufus G. Fellers, University of South Carolina; Eddie R. Fowler, Kansas State University; Joseph Frank, New Jersey Institute of Technology; Gaetano Giudice, Queensborough Community College; Paul E. Gray, University of Wisconsin - Platteville; Robert D. Hatch, Lawrence Institute of Technology; Michael F. Kavanaugh, Lake Superior State College; William J. Kerwin, The University of Arizona; Robert J. Krueger, The University of Wisconsin - Milwaukee; Edwin C. Lowenberg, Northern Arizona University; Charles G. Nelson, California State University, Sacramento; Edgar A. O'Hair, Texas Tech University; John O'Malley, University of Florida; Isaac R. Porche, Southern University; Arifur Rahman, University of South Alabama; James G. Simes, California State University, Sacramento; Charles E. Smith, The University of Mississippi; Rodney J. Soukup, University of Nebraska - Lincoln; David B. Taylor, Coastal Carolina Community College; Stephen P. Tubbs, The Pennsylvania State University - McKeesport; N. L. Weinberg, University of Miami

ENGINEERING PRESS, INC.   SAN JOSE, CALIFORNIA 95103-0001

Donald G. Newnan, Ph.D.
EXAM FILE Series Editor

©Copyright 1986, Engineering Press, Inc.
All rights reserved. Reproduction or
translation of any part of this work beyond
that permitted by section 107 or 108 of the
1976 United States Copyright Act without the
permission of the copyright owner is
unlawful.

Printed in the United States of America

1 2 3 4 5

Library of Congress Cataloging-In-Publication Data

Main entry under title:

Circuit analysis exam file.

(Exam file series)
1. Electric circuit analysis—Examinations, questions,
etc. I. Davis, Artice M., date.   II. Aronhime,
Peter. III. Series.
TK454.C627   1986   621.3815'3'076   85-25346
ISBN 0-910554-53-6

ENGINEERING PRESS, Inc.   P.O. Box 1   San Jose, California 95103-0001

Contents

Foreword	v
1. RESISTIVE CIRCUITS	
Definitions, Units, and Basic Concepts	1
Basic Circuit Laws	9
Analysis of Simple Circuits	15
Dependent Sources	31
2. GENERAL ANALYSIS METHODS	
Nodal Analysis	41
Mesh Analysis	59
3. NETWORK THEOREMS	
Linearity and Superposition	73
The Substitution Theorem	89
Thevenin's and Norton's Theorems	90
Maximum Power Transfer	115
4. NETWORK GRAPHS	
The Graph of a Circuit	123
Trees, Cotrees, and Cutsets	126
Independent Voltage Equations and Fundamental Cutsets	130
Independent Current Equations and Fundamental Loops	131
5. TRANSIENT ANALYSIS	
Inductors and Capacitors	133
Single Time Constant Circuits	143
Second Order Circuits	161

6. SINUSOIDAL ANALYSIS
Sinusoids, Phasors, and Complex Algebra	171
Complex Impedance and Admittance	177
Phasor Equivalent Circuits	184
Network Theorems and Nodal / Mesh Analysis	187
AC Steady State Power	194
Three Phase Circuits	213
Frequency Response	218

7. COMPLEX FREQUENCY
Poles, Zeros, and Network Functions	229
Natural and Forced Responses Derived from the Network Function	236

8. TWO PORT NETWORKS
Definition of a Two-Port; Port Conditions	241
Two-Port Parameters	242
Transformers	252

9. STATE VARIABLE ANALYSIS
Solution of the Vector-Matrix State Equation	265

10. FOURIER METHODS
The Trigonometric Fourier Series	269
Symmetry Properties	276
Response to Periodic Excitations	279
Complex Form of the Fourier Series	282
The Fourier Transform	285

11. LAPLACE TRANSFORMS
Impulse Response and Convolution	287
Definition of the LaPlace Transform and the System Function	292
Basic Transforms and Properties	295
Inverse Transforms by Partial Fraction Expansion	298
The S Domain Equivalent Circuit	305
The Complete Response of Circuits	308

Foreword

Many students find that previously-used examination problems are a useful supplement to their textbooks and lecture notes. Although copies of such examinations are often made available by professors, there is usually a need for more of this material than can be easily supplied by one teacher. Further, that material which *is* made available to a particular class reaches only a small fraction of the students in the country who would find it useful. Thus, the many, well-considered problems residing in professors' files across the country are unavailable to the people who could best use them. The purpose of this book is to make some of this material available.

Professors from around the country have agreed to open their exam files and to allow their problems and solutions to be published. These professors all teach an introductory course in circuit theory and use one of the three or four popular textbooks. The problems for publication were carefully selected to cover the fundamentals of circuits found in these textbooks. In general, the problems are just as they appeared in an actual course examination. Similarly, each solution is that prepared by the professor who wrote the exam problem. Thus this book is a combination of authentic examination problems, together with the professors' own solutions.

Each contributor has been careful to eliminate errors, so we hope there are none and expect but a few. If you find any, a note, mailed to the Engineering Press address, would be appreciated. I hope you find this material helps improve both your understanding of circuit theory and your examination scores.

Artice M. Davis
Editor

EXAM FILES

Professors around the country have opened their exam files and revealed their examination problems and solutions. These are actual exam problems with the complete solutions prepared by the same professors who wrote the problems. EXAM FILES are currently available for these topics:

- Circuit Analysis
- Dynamics
- Engineering Economy
- Fluid Mechanics
- Materials Science
- Mechanics of Materials
- Physics I Mechanics
- Physics II Heat, Light and Sound
- Physics III Electricity and Magnetism
- Probability and Statistics
- Statics
- Thermodynamics

The EXAM FILE series also includes three engineering license review books:

- Engineer-In-Training Exam File
- Civil Engineering License Exam File
- Mechanical Engineering License Exam File

Other EXAM FILES planned for release in the near future are:

- Chemistry
- Calculus I
- Calculus II
- Calculus III

For a description of all available EXAM FILES, or to order them, ask at your college or technical bookstore, or write to:

Engineering Press, Inc.
P.O. Box 1
San Jose, California 95103-0001

1
RESISTIVE CIRCUITS

DEFINITIONS, UNITS, AND BASIC CONCEPTS

━━ 1-1

A current of 5 μA flows through a wire.
  a) How many Coulombs of charge have passed through the wire in 10 seconds?
  b) How many Coulombs would flow through the wire in 2 years?

****

$$I = 5 \mu A = 5 \times 10^{-6} A$$

a) $t = 10$ seconds

$$Q = I \cdot t = (5 \times 10^{-6})(10) = \underline{5 \times 10^{-5} \text{ Coulombs}}$$

b) $t = 2 \text{ yr} \times \dfrac{365 \text{ day}}{1 \text{ yr}} \times \dfrac{24 \text{ hr}}{1 \text{ day}} \times \dfrac{60 \text{ min}}{1 \text{ hr}} \times \dfrac{60 \text{ sec}}{1 \text{ min}}$

$$= 6.31 \times 10^7 \text{ seconds}$$

$$Q = I \cdot t = (5 \times 10^{-6})(6.31 \times 10^7) = \underline{315 \text{ Coulombs}}$$

1-2

For the arrangement of voltage sources shown, find V_{AD}, V_{CD}, V_B, V_C, and V_A.

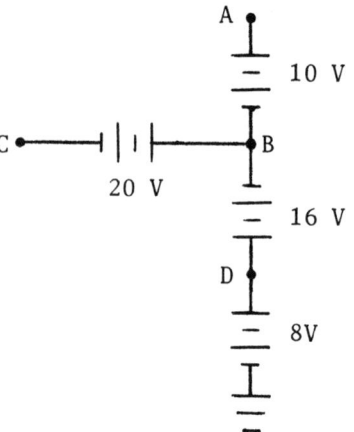

***

To find a voltage difference V_{XY}, trace circuit from X to Y and sum the voltages algebraically by affixing to each voltage the first polarity of each voltage difference encountered:

$$V_{AD} = 10V - 16V = \underline{-6V}.$$

$$V_{CD} = -20V - 16V = \underline{-36V}.$$

To find voltages with respect to ground, trace circuit from point to ground:

$$V_B = -16V + 8V = \underline{-8V}.$$

$$V_C = -20V - 16V + 8V \quad (= -20V + V_B)$$
$$= -36V + 8V = \underline{-28V}.$$

$$V_A = 10V - 16V + 8V \quad (= 10V + V_B)$$
$$= 18V - 16V = \underline{2V}.$$

Definitions, Units, And Basic Concepts / 3

1-3

An automobile battery is charged with a constant current of two Amps for five hours. Its terminal voltage is observed to be (11 + 0.5 t) volts where t is in hours.

a) Find the number of coulombs and the number of electrons which pass through the battery from t = 0 to t = 5 hrs.
b) Find the number of Joules delivered to the battery from t = 0 to t = 5 hrs.
c) If electricity costs 10¢/kwh, how much does the energy of part b cost?

****

a) $Q = \int_0^T I\,dt = \int_0^{5 \times 3600} 2\,dt$

$= \boxed{36000 \text{ Coulombs}}$ (1 hour = 3600 sec)

number of electrons $= 36000 \text{ Coulombs} \times \dfrac{1}{1.601 \times 10^{-19}}$

$= \boxed{2.25 \times 10^{23} \text{ electrons}}$

b) $W = \int_0^T p(t)\,dt = \int_0^{5 \times 3600} (2)\left(11 + \dfrac{.5t}{3600}\right)dt$

$= \left. 22t + \dfrac{2}{2} \cdot \dfrac{.5 t^2}{3600} \right|_0^{18000}$

$= 22 \times 18000 + \dfrac{.5(18000)^2}{3600} - 0$

$= \boxed{441,000 \text{ Joules}} = 441,000 \text{ Watt-sec}$

c) Cost $= 441,000\,J \times 10^{-3}\dfrac{kJ}{J} \times \dfrac{1}{3600}\dfrac{hr}{sec} \times \dfrac{10¢}{kwh}$

$= \boxed{1.225 ¢}$

1-4

When the current i is applied to the electric network shown, the voltage waveform v is produced. Determine expressions for the instantaneous power, p(t), and the energy, w(t), delivered to the network. Sketch p(t) and w(t).

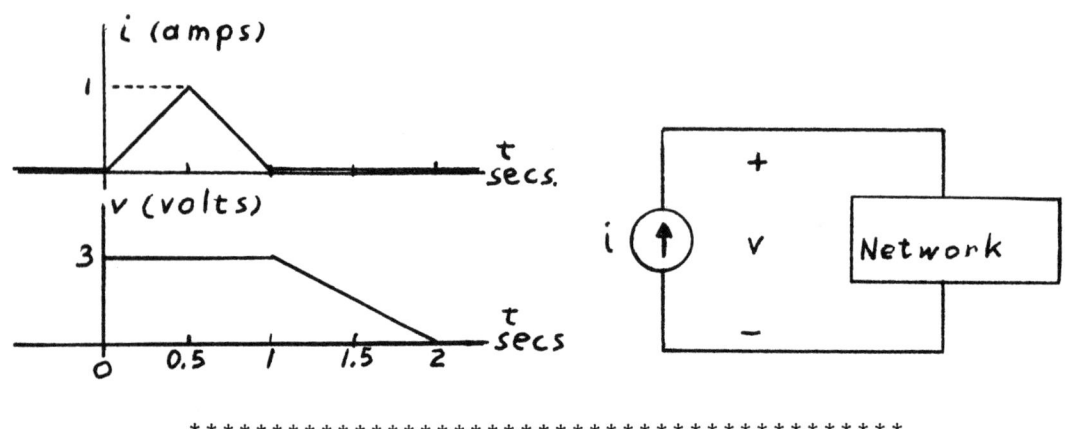

* *

TIME INTERVAL (SECS.)	CURRENT i (AMPS.)	VOLTAGE v (VOLTS)	$P(t) = vi$ (WATTS)
$0 \leq t \leq 0.5$	$2t$	3	$6t$
$0.5 \leq t \leq 1.0$	$-2t+2$	3	$-6t+6$
$1.0 \leq t \leq 2.0$	0	$-3t+6$	0
$2.0 \leq t$	0	0	0

$$w(t) = \int_{-\infty}^{t} P(t)\, dt$$

For $0 \leq t \leq 0.5$ sec., $w(t) = \int_{-\infty}^{0} P\,dt + \int_{0}^{t} 6t\, dt = 3t^2$ JOULES

At $t = 0.5$ sec., $w(t) = 0.75$ JOULES

For $0.5 \leq t \leq 1.0$ sec., $w(t) = \int_{-\infty}^{0} P\,dt + \int_{0}^{0.5} P\,dt + \int_{0.5}^{t} (-6t+6)\,dt$

$$= -3t^2 + 6t - 1.5 \text{ JOULES}$$

For $t \geq 1.0$ sec., $w(t)$ NEITHER INCREASES NOR DECREASES SINCE $P(t) = 0$.

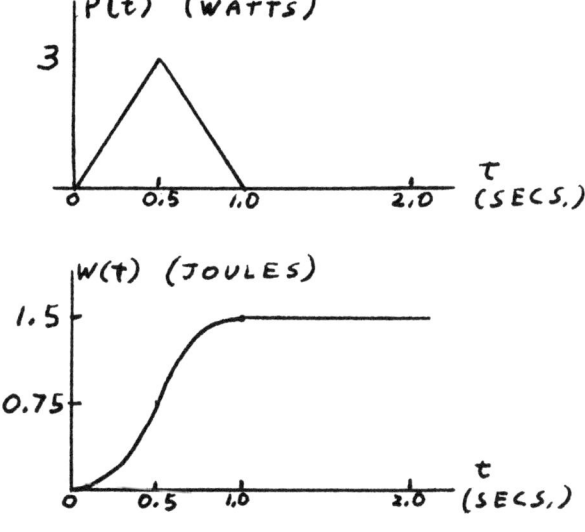

1-5

How much current is flowing in a conductor if 24 coulombs of charge pass a given point in the conductor in 0.5 seconds?

****

By definition, current is the rate of flow of charge, i.e., $I = \dfrac{Q}{t}$

$\therefore \quad I = \dfrac{Q}{t}$

$I = \dfrac{24 \text{ coul.}}{0.5 \text{ sec.}} = \underline{48 \text{ amperes}}$

1-6

For the device shown below:
$$i(t) = 20(1 - e^{-2t}) \text{ amps}$$
$$v(t) = 10 \sin(4t) \text{ volts}$$

Find: a.) the charge, $q(t)$, in the device at $t = 2$ sec.
$q(0) = 0$ coulombs

b.) the power supplied to the device at $t = 2$ sec.

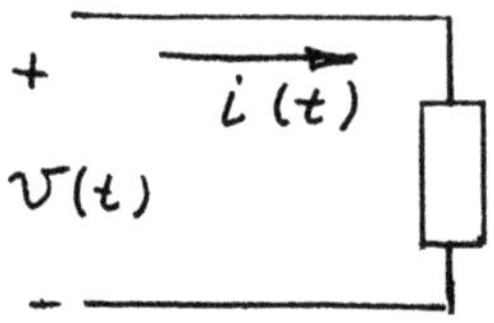

***

a) SINCE $i = \dfrac{dq}{dt}$

$$q = \int_0^T i(\tau)\,d\tau + q(0)$$

$$q = \int_0^2 20(1 - e^{-2\tau})\,d\tau + 0$$

$$q = 20\left[\tau - \dfrac{e^{-2\tau}}{-2}\right]_0^2$$

$$q = 30.2 \text{ COUL.} \qquad\qquad \text{ANS a}$$

b) $p(t) = v(t)\,i(t)$

$$= 10 \sin(8)\,20(1 - e^{-4})$$

$$= 194.3 \text{ WATTS} \qquad\qquad \text{ANS b}$$

1-7

What is the absorbed power and delivered power for each of these elements?

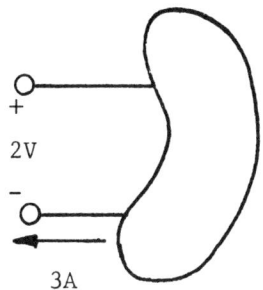

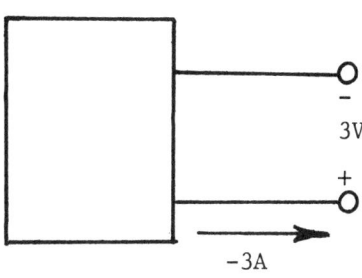

****

A) Power Absorbed: $P_A = VI = 2 \cdot 3 = 6\,W.$
   Power Delivered: $P_D = -VI = -2 \cdot 3 = -6\,W.$

B) Power Absorbed: $P_A = -VI = -(3 \cdot (-3)) = 9\,W.$
   Power Delivered: $P_D = VI = 3 \cdot (-3) = -9\,W.$

1-8

A component, COM, exhibits a current behavior, IM, and a voltage behavior, VM, as shown. Find the value of the power supplied by COM and the equivalent resistance, RB, attached to the source.

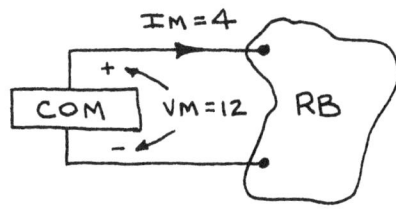

****

The product of the two source behaviors is (4 * 12 = 48 watts) supplied by COM. The ratio of the two behaviors is RB = 12/4 = 3 ohms.

1-9

For the resistive network shown below, find the equivalent resistance, R_{eq}, in ohms. All resistance values are in ohms on the diagram.

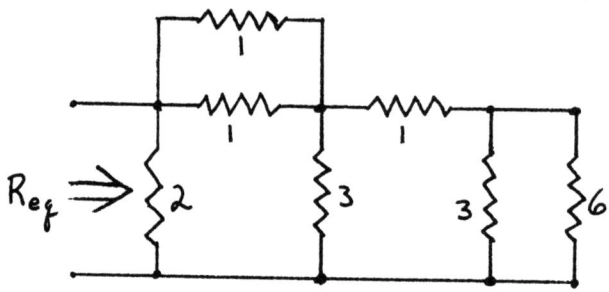

****

The parallel combination of 3 and 6 ohms becomes:

$$R_{36} = \frac{3 \times 6}{3+6} = 2 \, \Omega$$

The parallel combination of the two 1Ω resistors becomes: $R_{11} = \frac{1 \times 1}{1+1} = 0.5 \, \Omega$

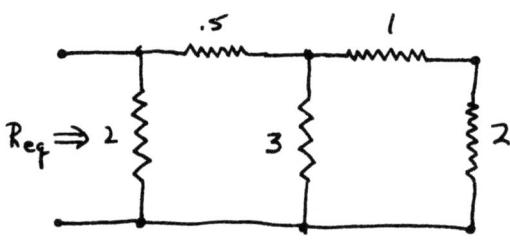

The 1 and 2 ohms in series becomes 3Ω and in parallel with 3Ω becomes

$$R_{33} = \frac{3 \times 3}{3+3} = 1.5 \, \Omega$$

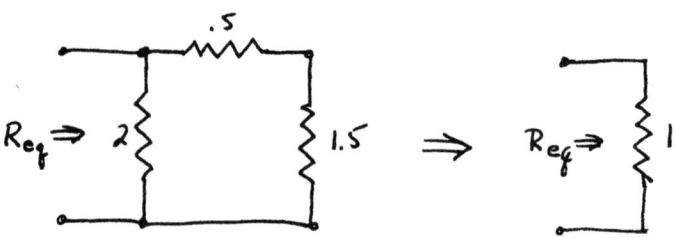

The 0.5 and 1.5 ohms in series become 2Ω and in parallel with 2Ω becomes, $R_{eq} = 1 \, \Omega$.

1-10

A component, COM, exhibits a voltage behavior, VX, and a current behavior, IX, as shown. Find the resistance and the power absorbed by the COM.

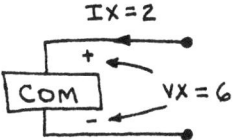

*****

Using the voltage drop convention the ratio of the behaviors is the resistance; $\left\{\frac{VX}{IX}\right\} = \frac{6}{2} = 3$ ohms. Using the power Absorbing convention the product of the behaviors is the power absorbed; $\{VX\}\{IX\} = 6*2 = 12$ watts.

BASIC CIRCUIT LAWS

1-11

a) Find all unknown voltages and currents.
b) Find the power flow <u>into</u> each element.

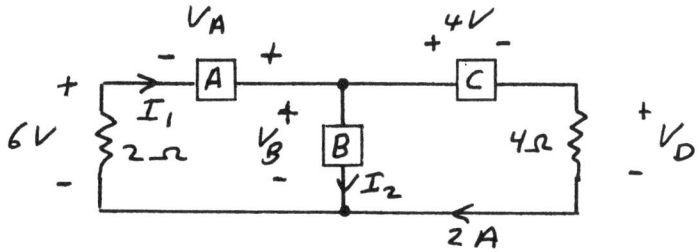

*****

a)
$I_1 = -6/2 = \boxed{-3A}$

$I_2 = I_1 - 2 = \boxed{-5A}$

$V_D = 2 \times 4 = \boxed{8V}$

$V_B = V_D + 4 = \boxed{12V}$

$V_A = V_B - 6 = \boxed{6V}$

b)
$P_{2\Omega} = (-3)^2 \times 2 = \boxed{18W}$

$P_{4\Omega} = (2)^2 \times 4 = \boxed{16W}$

$P_A = -V_A I_1 = -(6)(-3) = \boxed{18W}$

$P_B = V_B I_2 = (12)(-5) = \boxed{-60W}$

$P_C = 4 \times 2 = \boxed{8W}$

Check that power adds to 0
$18 + 16 + 18 - 60 + 8 = 0$ ✓

1-12

Find V the voltage between points 1 and 2.

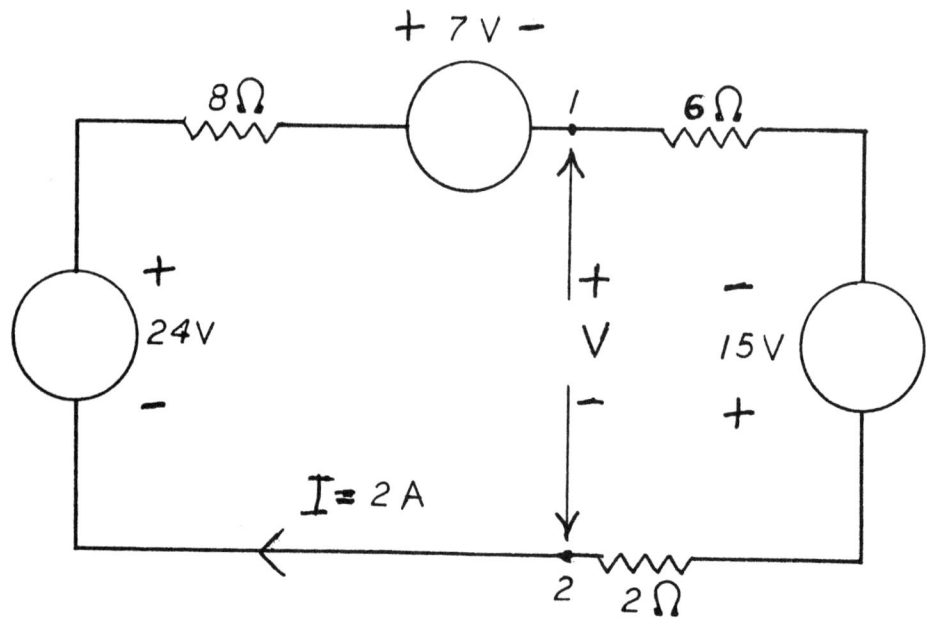

Kirchoff's voltage law around section of the circuit to the left of points 1 and 2

$$-24 + 2 \times 8 + 7 + V = 0$$

$$V = 24 - 16 - 7 = 1 \text{ Volt}$$

Sum voltages around section of circuit to the right of points ① and ②

$$-V + 2 \times 6 - 15 + 2 \times 2 = -1 + 12 - 15 + 4 = 0$$

Check

1-13

Calculate the terminal voltage and current of the current source for the given circuit.

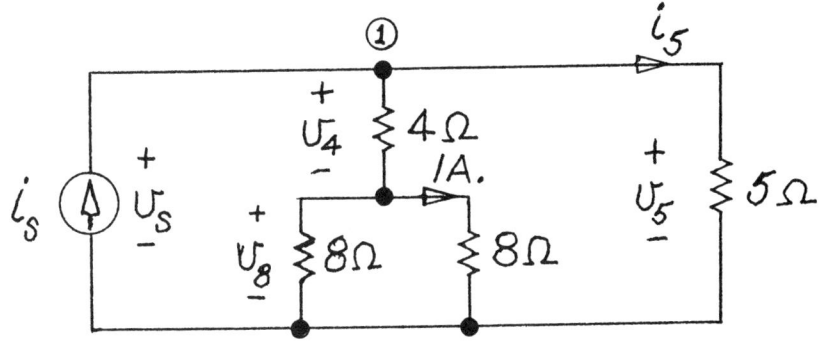

****

$v_8 = 8V$. BECAUSE BOTH 8Ω RESISTORS ARE IN PARALLEL.

$i_4 = 2A$. FROM OHM'S LAW, BOTH 8Ω RESISTORS HAVE $1A$. AND KCL YIELDS $2A$. THRU 4Ω.

$v_4 = 8V$. FROM OHM'S LAW.

$v_5 = 16V$. FROM KVL AROUND $8\Omega, 4\Omega, \& 5\Omega$ LOOP.

$i_5 = 3.2A$. FROM OHM'S LAW.

$\underline{v_s = 16V.}$ FROM KVL AROUND $i_s \& 5\Omega$ LOOP.

$\underline{i_s = 5.2A.}$ FROM KCL AT NODE 1.

1-14

For the network shown, calculate the power absorbed or generated by each branch.

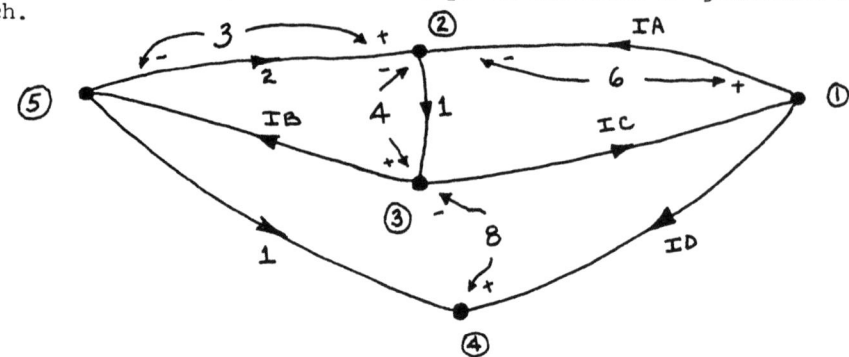

branch between node-pair	branch current	node-pair voltage	power absorbed	power generated
(1,2)	-1	6		6
(1,3)	2	2	4	
(1,4)	-1	-6	6	
(1,5)	none	9	none	none
(2,3)	1	-4		4
(2,4)	none	-12	none	none
(2,5)	-2	3		6
(3,4)	none	-8	none	none
(3,5)	3	7	21	
(4,5)	-1	15		15
			31 =	31
			checks	

1-15

A network has four nodes labeled 0, 1, 2, and 3. Calculate all the node-pair voltages.

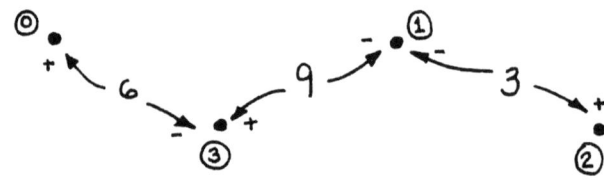

****

V(x,y) = voltage drop from x (assumed +) to y (assumed -)

The number of node-pair voltages required is the number of combinations $\frac{4!}{2!2!} = 6$. Use KVL around a node path --- six times; e.g., V(0,1) = V(0,3) + V(3,1) = 6 + 9 = 15. Thus:

$$V(0,1) = 15 \quad V(0,3) = 6 \quad V(3,1) = 9$$
$$V(0,2) = 12 \quad V(1,2) = -3 \quad V(3,2) = 6$$

1-16

In the following circuit determine the current in the 3 Ohm resistor, using Kirchhoff's Voltage and Current laws.

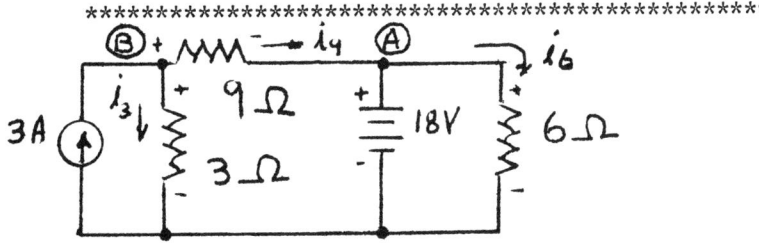

Assume directions for i_3 and i_9 as shown and then label the polarity of the voltages as shown:

KCL at node B yields:

$-i_3 - i_9 + 3 = 0$ (using entering the node as positive) 1)

Since the current in the 6 ohm resistor is independent of the circuitry to the left of the voltage source, skip KCL at node **A**. Apply KVL around a loop that does not contain a current source.

Using the inner loop:

$+v_3 - v_9 - 18 = 0$

Ohms Law applied to each resistor yields:

$3i_3 - 9i_9 = 18$ 2)

Solving simultaneously for i_3, $i_3 = 3.75$ A.

Note that by writing at node **A**, an extra unknown (the current through the battery would be incurred. The current thru the 6 Ohm resistor is 18/6=3A.

1-17

Determine the value of R_2 such that $V_O = 22$ volts. Will a 1/4 watt resistor be satisfactory?

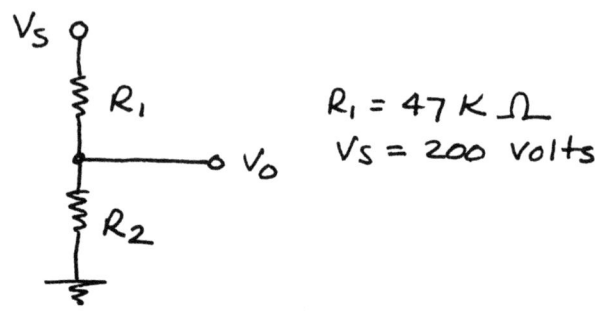

***

$$V_O = \frac{R_2 V_S}{R_1 + R_2} \implies R_2 = \frac{V_O R_1}{V_S - V_O} = \frac{(22)(47 \times 10^3)}{200 - 22}$$

$$\boxed{R_2 = 5808.99 \, \Omega}$$

$$P_{R_2} = \frac{V_{R_2}^2}{R_2} = \frac{V_O^2}{R_2} = \frac{(22)^2}{5808.99} = 83.33 \times 10^{-3} \text{ watts}$$

Since $P_{R_2} < 0.250$ watts, a 1/4 watt resistor will suffice for R_2.

1-18

A network has nine branch currents. Find the five unknown currents.

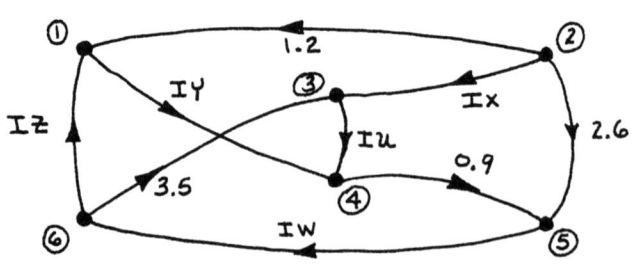

***

Use KCL at node ②: $I_X = -3.8$; then $I_W = 3.5$; $I_Z = 0$; $I_Y = 1.2$; $I_u = -0.3$

ANALYSIS OF SIMPLE CIRCUITS

━━ 1-19

For the circuit below, determine
a. the total resistance Rab
b. the current I
c. the voltage developed across R_2 by voltage division

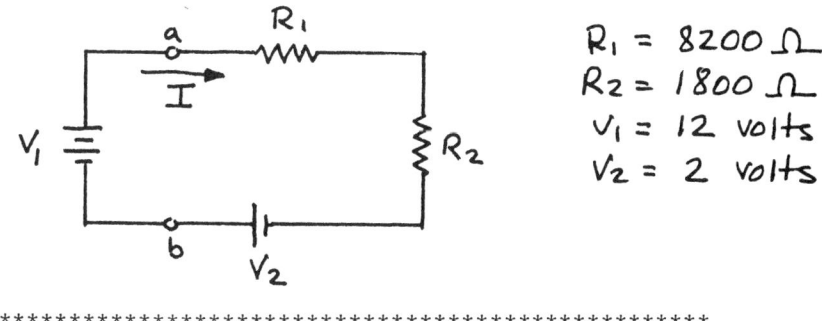

$R_1 = 8200\ \Omega$
$R_2 = 1800\ \Omega$
$V_1 = 12\ \text{volts}$
$V_2 = 2\ \text{volts}$

****

a. $R_{ab} = R_1 + R_2 = 8200 + 1800 = 10,000\ \Omega$

$$\boxed{R_{ab} = 10\ K\Omega}$$

b. $V_1 = IR_1 + IR_2 - V_2$

$I = \dfrac{V_1 + V_2}{R_1 + R_2} = \dfrac{14v}{10K}$; $\boxed{I = 1.4\ mA}$

c. $V_{R_2} = \dfrac{R_2(V_1 + V_2)}{R_1 + R_2} = \dfrac{1800\,(14v)}{10K} = 2.52\ \text{volts}$

$$\boxed{V_{R_2} = 2.52\ \text{volts}}$$

1-20

Two voltage sources are available that are labeled v_1 and v_2. Devise a resistor network that will yield

$$v_o = \frac{2}{5} v_1 + \frac{1}{5} v_2.$$

Use the minimum number of resistors and assume no load will be placed on the network.

* *

TRY TWO RESISTORS AS SHOWN:

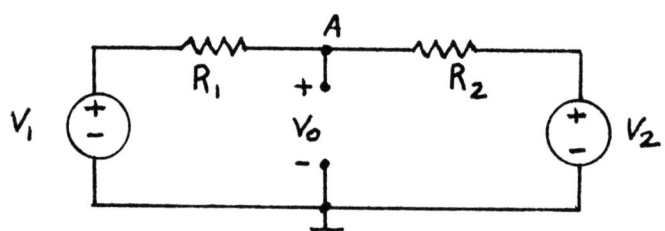

NODE EQUATION AT A:

$$\frac{V_o - V_1}{R_1} + \frac{V_o - V_2}{R_2} = 0$$

$$V_o = \frac{V_1 R_2}{R_1 + R_2} + \frac{V_2 R_1}{R_1 + R_2} = \frac{2}{5} V_1 + \frac{1}{5} V_2$$

$$\frac{R_2}{R_1 + R_2} = \frac{2}{5} \quad \text{AND} \quad \frac{R_1}{R_1 + R_2} = \frac{1}{5}$$

$$R_1 = \frac{5}{3}\ \Omega \qquad\qquad R_2 = \frac{10}{3}\ \Omega$$

1-21

Simplify the circuit by reducing it to one equivalent resistance across A-B. What is the power supplied by the 12V source?

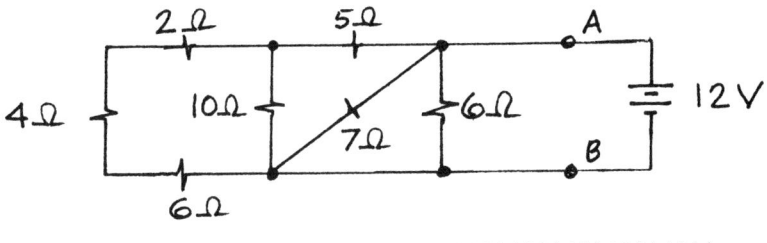

****

ADD 3 SERIES RESISTANCES $2+4+6=12\,\Omega$

EQUIVALENT RESISTANCE FOR $12\,\Omega$ AND $10\,\Omega$ IN PARALLEL: $R_{e_1} = \frac{10 \times 12}{10+12} = 5.45\,\Omega$

R_{e_1} IS IN SERIES WITH $5\,\Omega$ $R_{e_2} = R_{e_1} + 5 = 10.45$

R_{e_2} IS IN PARALLEL WITH THE $7\,\Omega$ AND $6\,\Omega$

$$\frac{1}{R_E} = \frac{1}{10.45} + \frac{1}{7} + \frac{1}{6} = 0.41$$

$$R_E = 2.47\,\Omega \quad ; \quad P = \frac{V^2}{R_E} = \frac{(12)^2}{2.47} = 58.3\,W$$

1-22

Calculate the total resistance seen by the source in the following circuit.

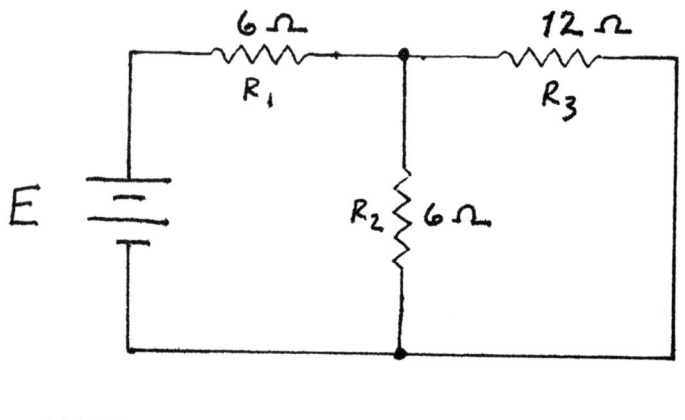

***

R_2 and R_3 are in parallel with each other and may be replaced with an equivalent resistance.

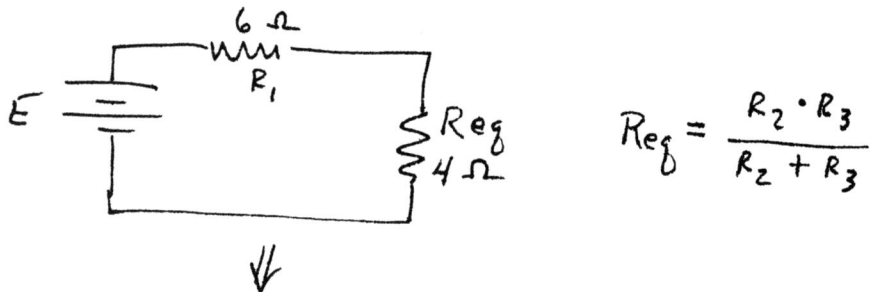

$$R_{eq} = \frac{R_2 \cdot R_3}{R_2 + R_3}$$

R_1 and the equivalent resistance are in series.

$R_T = R_1 + R_{eq}$

$R_T = 10\,\Omega$

1-23

Determine the equivalent resistance to the right of terminals a - b. Calculate v and i.

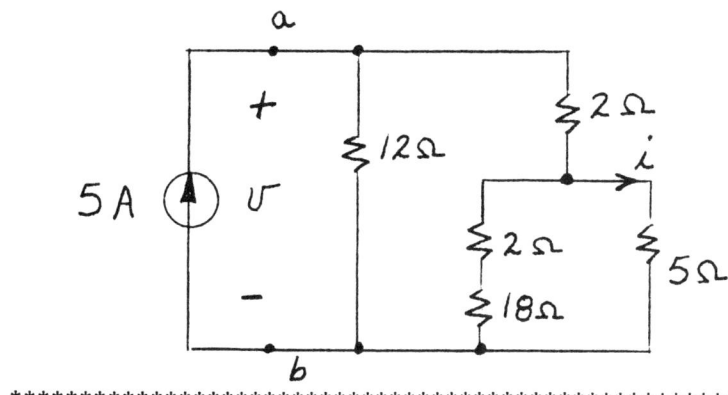

THE EQUIVALENT RESISTANCE OF THE RIGHT HAND BRANCH IS $2 + 5\|(18+2) = 6\,\Omega$. THEREFORE, AT a-b WE SEE $6\|12 = 4\,\Omega = R_{EQ}$.

THIS RESULTS IN $v = 5(R_{EQ}) = 20\,V$, AND $i = \dfrac{12}{12+6} \times \dfrac{20}{20+5} \times 5 = \dfrac{2}{3} \times \dfrac{4}{5} \times 5 = \dfrac{8}{3}\,A$.

1-24

Known voltages are $\begin{bmatrix} V_{AC} \\ V_{BC} \\ V_{DC} \end{bmatrix} = \begin{bmatrix} -0.750 \\ 3.500 \\ -3.125 \end{bmatrix}$

Calculate the equivalent resistance seen by the current source and the total power absorbed.

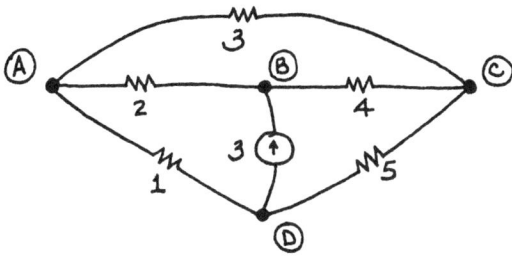

R_{equiv} = ratio of source voltage and current = $\dfrac{V_{BC} - V_{DC}}{3} = \dfrac{3.5 - (-3.125)}{3} = \dfrac{6.625}{3} = 2.208$ ohms. Pwr absorbed equals pwr supplied = $3 * V_{BD} = 3 * 6.625 = 19.88$ watts.

1-25

The power to resistance R_1 in the circuit shown is 9 W; find the current through each resistance in mA and the source voltage V.

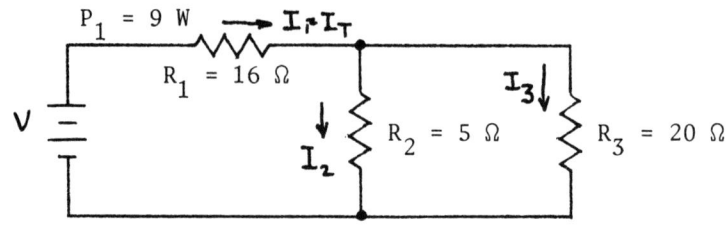

****

1) $I_T = I_1 = \sqrt{\dfrac{P_1}{R_1}} = \sqrt{\dfrac{9}{16}} = .75\,A = \underline{750\,mA}$.

2) $R_{23} = R_2 \| R_3 = \dfrac{5\Omega \cdot 20\Omega}{5\Omega + 20\Omega} = \dfrac{100\Omega}{25} = 4\Omega$.

3) $I_2 = I_T \dfrac{R_{23}}{R_2} = 750\,mA \cdot \dfrac{4\Omega}{5\Omega} = \underline{600\,mA}$.

$I_3 = I_T \dfrac{R_{23}}{R_3} = 750\,mA \cdot \dfrac{4\Omega}{20\Omega} = \underline{150\,mA}$.

Check: $I_T = I_2 + I_3 = 600\,mA + 150\,mA = 750\,mA$.

4) $R_T = R_1 + R_{23} = 20\Omega$.

5) $V = I_T R_T$
   $= 750m \cdot 20 = \underline{15V}$.

1-26

Find V_{ab}, the voltage across the open circuit in the following circuit.

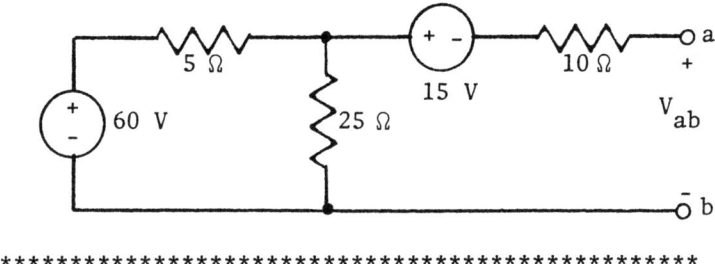

A good approach is to first determine the voltage drop V_{25}, top to bottom across the 25-Ω resistor, and then find V_{ab} by applying KVL to the right-hand mesh. Since no current can flow in the open circuit, the current is the same in the 5-Ω and 25-Ω resistors, which means that we can use voltage division to find V_{25}.

$$V_{25} = \frac{25}{5+25} \cdot 60 = 50 \text{ V}$$

Before we can apply KVL to the right-hand mesh, we need to find the voltage drop across the 10-Ω resistor. It is 0 V because the current through it is 0 A as a result of the open circuit. Finally, applying KVL to the right-hand mesh by summing voltage drops in a clockwise direction, we obtain

$$15 + 0 + V_{ab} - 50 = 0$$

So, $V_{ab} = 35$ V.

1-27

Find the value of the current I.

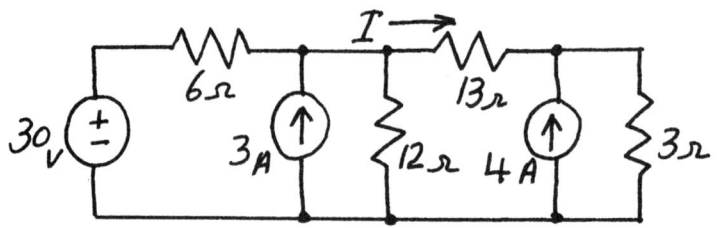

****

This problem is most easily done by a series of source conversions.

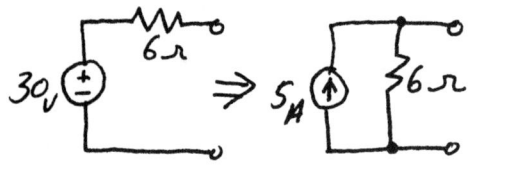

Combining the 5A and the 3A sources and the 6Ω || with the 12Ω

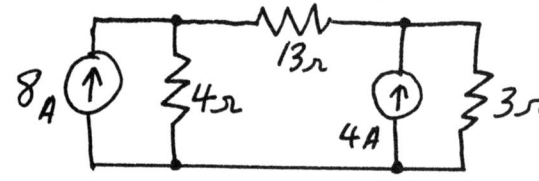

Converting both current sources to voltage sources we obtain

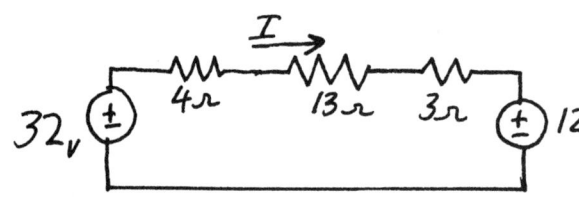

$$I = \frac{32-12}{20} = \underline{\underline{1\,A}}$$

1-28

In the following circuit find V_{ag}, which is referenced positively at node a and negatively at node g.

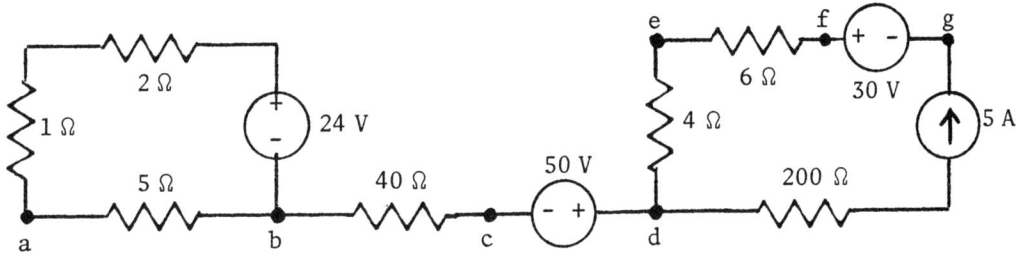

***

The key to this problem is to find a path from node a to node g in which all voltages are either known or can be easily determined. By inspection, such a path includes nodes a, b, c, d, e, f, and g. Note that this path does not include the current source, which has an unknown voltage. So by KVL,

$$V_{ag} = V_{ab} + V_{bc} + V_{cd} + V_{de} + V_{ef} + V_{fg}$$

Now determining the voltages, we find by voltage division that

$$V_{ab} = \frac{5}{5+1+2} 24 = 15 V$$

Also, $V_{bc} = 0 V$ because no current flows through the 40-Ω resistor, as is evident from enclosing either mesh in a closed surface and applying KCL. By inspection, $V_{cd} = -50V$. The 5-A current from the current source flows through the 4-Ω and 6-Ω resistors, producing $V_{de} = -20V$ and $V_{ef} = -30 V$. And by inspection, $V_{fg} = 30V$. So,

$$V_{ag} = 15 + 0 - 50 - 20 - 30 + 30 = -55 V$$

1-29

Determine the equivalent conductance of the circuit between terminals a & b.

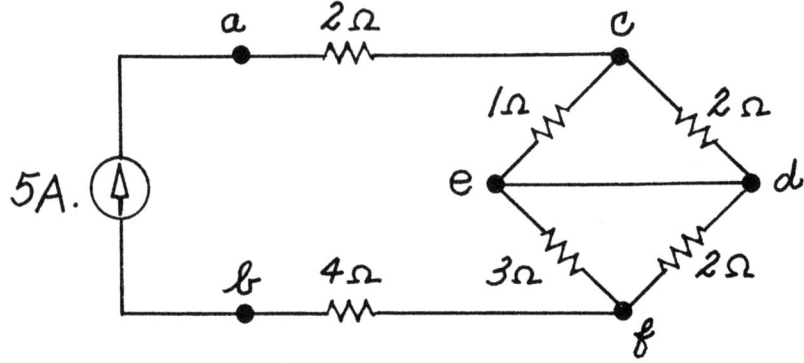

****

RESISTORS 1Ω & 2Ω ARE IN PARALLEL.
RESISTORS 3Ω & 2Ω ARE IN PARALLEL.

$$R_{ce} = R_{cd} = \frac{(1)(2)}{1+2} = 0.67\Omega \text{ & } R_{ef} = R_{df} = 1.2\Omega$$

$$R_{ab} = R_{ac} + R_{ce} + R_{ef} + R_{fb} = 7.87\Omega$$

$$G_{ab} = \frac{1}{R_{ab}} = \underline{0.127\ S.}$$

1-30

Find the current I_L that flows through the resistor R_L using Source Conversion.

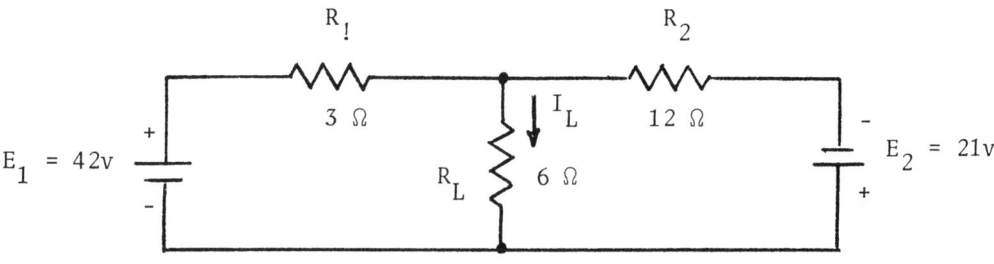

***

Convert E_1 and R_1 to a current source (I_1)
Convert E_2 and R_2 to a current source (I_2)

$$I_1 = \frac{E_1}{R_1} = \frac{42V}{3\Omega} = 14a$$

$$I_2 = \frac{E_2}{R_2} = \frac{21V}{12\Omega} = 1.75a$$

Combine the parallel current sources I_1 and I_2 (I_{12})
Combine the parallel resistors R_1 and R_2 (R_{12})

$$I_{12} = I_1(\downarrow) - I_2(\uparrow) = 14a - 1.75a = 12.25a \; (\downarrow)$$

$$R_{12} = R_1 \| R_2 = \frac{(3\Omega)(12\Omega)}{3\Omega + 12\Omega} = 2.4\Omega$$

Calculate I_L (current divider rule)

$$I_L = \frac{R_{12}}{R_{12} + R_L} \times I_{12} = \frac{2.4\Omega}{2.4\Omega + 6\Omega} \times 12.25a = 3.5a$$

1-31

Determine the value of R in the circuit.

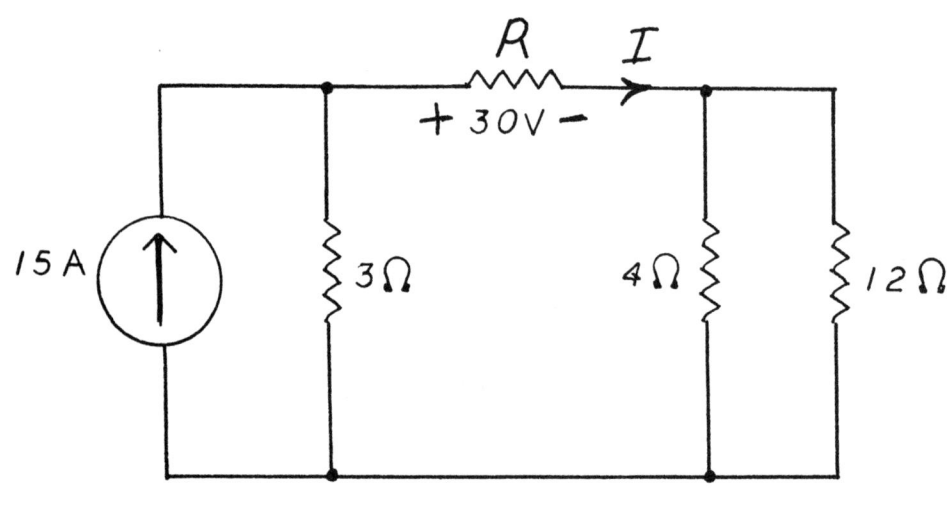

***

4 ohm and 12 ohm resistors are in parallel.

$$\frac{12 \times 4}{12+4} = 3 \text{ ohms}$$

This is in series with R which gives $(R+3)$ ohms.

By Current division:

$$I = 15 \times \frac{3}{(R+3)+3} = \frac{45}{R+6} \text{ Amps.}$$

Also by Ohm's Law:

$$I = \frac{30}{R} \text{ Amps.}$$

$$\frac{30}{R} = \frac{45}{R+6} \quad ; \quad 30R + 180 = 45R$$

$$15R = 180 \qquad \underline{R = 12 \text{ ohms.}}$$

1-32

Determine a single resistor R that is equivalent to the resistor network shown.

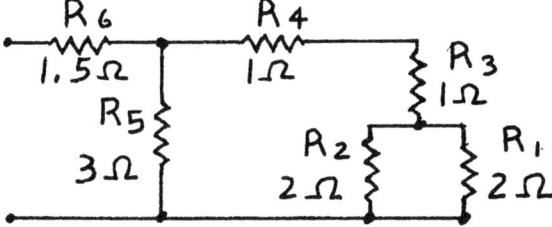

* *

$$R = \{[(R_1 \| R_2) + R_3 + R_4] \| R_5\} + R_6$$

$$R = \frac{[1+1+1]\,3}{3+3} + 1.5 = 3\,\Omega$$

1-33

Given that $R_{in} = 3$ ohms. What is R? (All resistors in ohms).

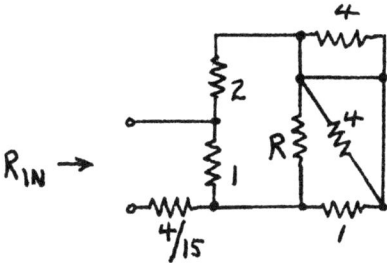

* *

REDUCING, THE 4 Ω RESISTORS ARE SHORTED:

THEN, $1 = \dfrac{4}{15} + \dfrac{(1)\,\dfrac{2+3R}{1+R}}{1 + \dfrac{2+3R}{1+R}}$

OR, $\dfrac{11}{15} = \dfrac{2+3R}{3+4R}$

SOLVING, $R = 3\,\Omega$

1-34

Determine the equivalent resistance of the circuit between terminals a & b.

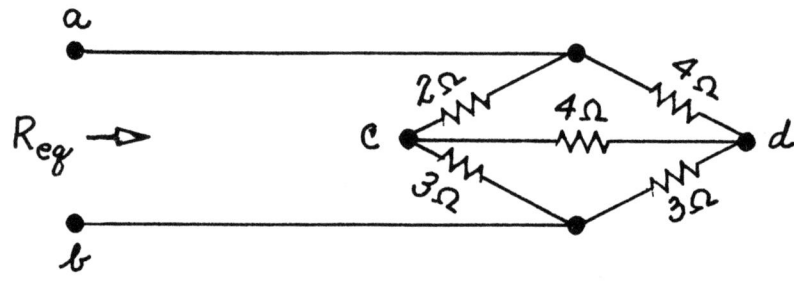

****

PERFORM A Δ TO Y TRANSFORMATION OF THE LOWER Δ.

$$R_1 = \frac{(3)(4)}{3+4+3} = 1.2 \, \Omega$$

$$R_2 = 1.2 \, \Omega$$

$$R_3 = \frac{(3)(3)}{3+4+3} = 0.9 \, \Omega$$

$(2+R_1) \| (4+R_2) = 1.98 \, \Omega$ & $\underline{R_{eq} = 2.88 \, \Omega}$

1-35

Find the power dissipated in each resistance and the power supplied by each source. Show the sum of the powers is zero.

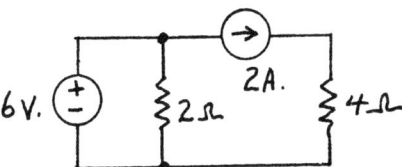

****

$P_{4\Omega} = I^2R = 2^2(4) = 16\text{ W.}$

$P_{2\Omega} = \dfrac{V^2}{R} = \dfrac{6^2}{2} = 18\text{ W.}$

$P_{6V} = V_6 I_6 = 6(-5) = -30\text{ W.}$

$P_{2A} = V_{2A}(2) = -2(2) = -4\text{ W.}$

$I_6 = 2 + \dfrac{6}{2} = 5\text{A.}$

KVL: $V_{2A} + 8 - 6 = 0$, $V_{2A} = -2$

SUM OF POWERS = POWER IN R'S + POWER IN SOURCES

SUM = $(16 + 18) + (-30 - 4) = 0\text{ W.}$

1-36

Note switch closure at three seconds. Placing an ohmmeter at node-pair (A, D) calculate its reading at times $t = 1$ and $t = 4$ seconds.

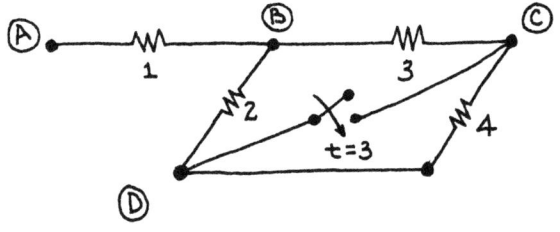

****

at $t = 1$:

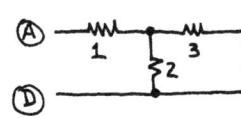

$R_{eq} = 1 + 2\|7 = 1 + \dfrac{14}{9} = \dfrac{23}{9}$ ohms

at $t = 4$:

$R_{eq} = 1 + 2\|3 = 1 + \dfrac{6}{5} = \dfrac{11}{5}$ ohms

1-37

Find output voltages V_{BA} and V_{CA} using the voltage divider rule.

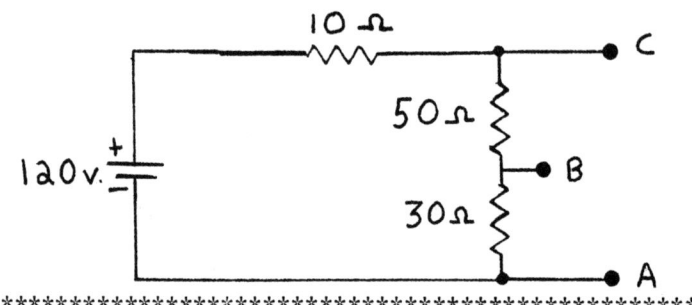

****

$$V_{BA} = 120 \frac{30}{30+50+10} \qquad V_{CA} = 120 \frac{50+30}{30+50+10}$$

$$= \underline{40 \text{ VOLTS}} \qquad\qquad = \underline{106.7 \text{ VOLTS}}$$

also

$$V_{CB} = 120 \frac{50}{30+50+10} = \frac{200}{3} \text{ Volts}$$

$$V_{AB} = -120 \frac{30}{30+50+10} = -40 \text{ volts}$$

1-38

Determine the voltage drop V_{AO} shown on the diagram.

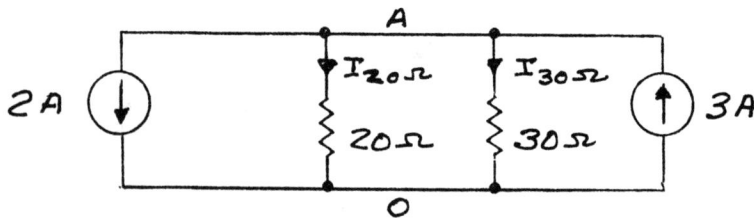

****

KCL Node A $\overset{(\sum_A I_{out} = 0)}{}$ $+2 + I_{20\Omega} + I_{30\Omega} - 3 = 0$

By Ohm's law: $\qquad +2 + \frac{1}{20} V_{AO} + \frac{1}{30} V_{AO} - 3 = 0$

Solving for V_{AO}: $\qquad V_{AO} = \left(\frac{60}{5}\right)(+1) = +12 V$

As a check: $\qquad +2 + \frac{12}{20} + \frac{12}{30} - 3 = 0$

DEPENDENT SOURCES

1-39

Given the following circuit containing a dependent source, find the power absorbed by the 100 Ω resistor and the dependent source.

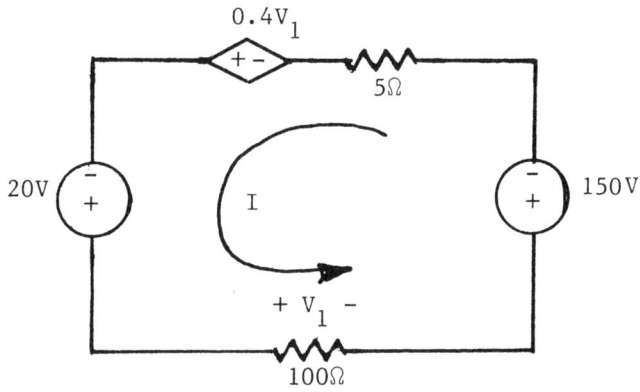

****

By Kirchhoff's Voltage Law, Equations for This Circuit Are

KVL: $20 + 0.4V_1 - 5I - 150 - 100I = 0$

Auxiliary EQ: $V_1 = 100I$

These Equations can Be Solved by Back Substitution As

$$20 + 0.4(100I) - 5I - 150 - 100I = 0$$
$$-130 - 65I = 0$$
$$I = -130/65 = -2 \text{ A}.$$

The Power Absorbed Can Now be Calculated As

$$P_{100} = I^2 R = (-2)^2 \cdot 100 = 400 \text{ W}.$$

$$P_{Dep.\ Source} = -VI = -0.4(100 \cdot (-2)) \cdot (-2) = -160 \text{ W}.$$

1-40

The switch opens at five seconds. For (t < 5) the box absorbs 150 watts. For (t > 5) calculate the value of RX required such that the box absorbs 65 watts.

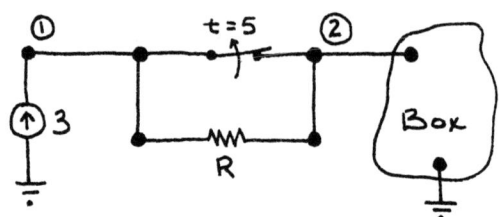

Notice that 3 Amps is supplied to the box in both the open and closed switch positions. For the power absorbed to vary the box must contain a controlled source (e.g., a VCVS) and a series resistance, RBOX.

For t<5   V(1,2)=0 and
$$R_{BOX} = \frac{150 \text{ watts}}{3^2} = 16.67 \text{ ohms}.$$

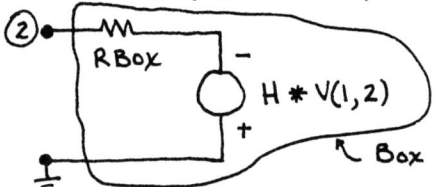

For t>5   V(1,2) = 3*R   and

RBOX absorbs 150 watts but the box absorbs 65 watts. Therefore the VCVS must supply (150-65 = 85 watts);
$$85 = 3\{H * V(1,2)\} = 9 * H * R. \text{ Also } V1 = 3R + 21.67; R \neq 0.$$
The problem has many answers depending upon your design requirements for gain H and resistance R; e.g., if H = 3 then R = 3.15.

1-41

Determine v_o for the network shown. The network contains an independent voltage source, v_1, and a current-controlled-current-source (ICIS), Ki_x.

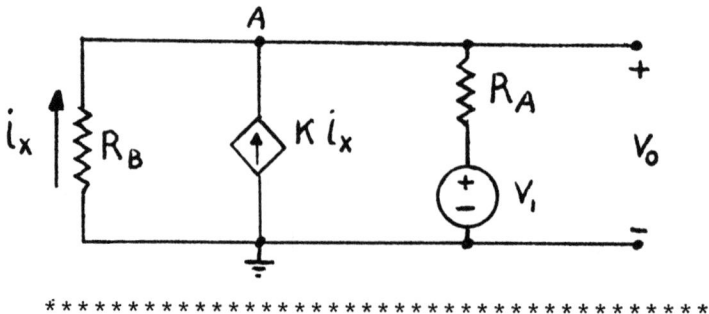

NODE A: $\quad -i_x - Ki_x + \dfrac{V_o - V_1}{R_A} = 0$

AUXILIARY EQUATION: $\quad V_o = -i_x R_B$
(USE TO ELIMINATE i_x)

$$V_o = V_1 \dfrac{R_B}{R_A(1+K) + R_B}$$

1-42

Find the power in watts supplied by the dependent source in the following circuit.

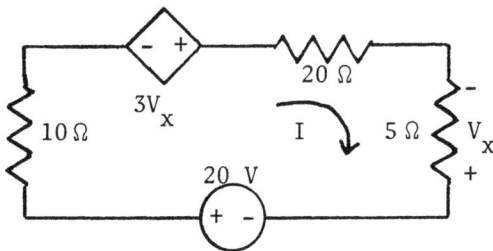

***

By inspection, the controlling voltage is $V_x = -5I$, which means that there is a voltage drop of $-3V_x = 15I$ across the dependent source in the direction of I. Applying KVL, we obtain

$$(10 + 15 + 20 + 5)I = 20$$

and so $\quad I = \dfrac{20}{50} = 0.4 \text{ A}$

Consequently, there is a voltage drop across the dependent source of $15(0.4) = 6V$, positive on the left-hand terminal. Since the 0.4-A current enters this positive terminal, the dependent source absorbs

$$P = VI = 6(0.4) = 2.4 \text{ W}$$

and so supplies -2.4 W.

1-43

In this circuit, determine I if α = 3.

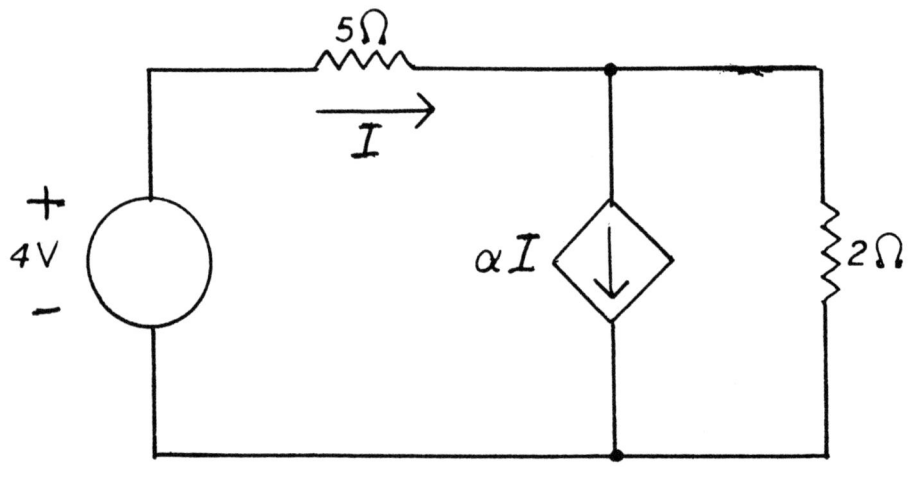

***

Current through the 2 ohm resistor is $I - \alpha I$.

Kirchoff's voltage law around outer loop. $(\sum_{\text{loop}} V_{\text{drops}} = 0)$

$$-4 + 5I + (1-\alpha)I \times 2 = 0 \qquad \alpha = 3$$

$$-4 + 5I - 2 \times 2 I = 0$$

$$5I - 4I = 4$$

$$\underline{I = 4 \text{ Amperes.}}$$

1-44

A. What type of controlled source do we have in the circuit shown?
B. What total resistance appears between the terminals?

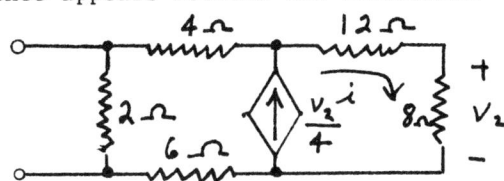

****

A. Since the source current depends upon a voltage elsewhere in the circuit, it is a "voltage-controlled current source"

B. By Kirchoff's current law, the same current flows through the 12Ω and the 8Ω. $V_2 = 8i_2$ and the voltage across the 12Ω is $12i_2 = 12V_2/8 = 1.5V_2$

There fore, by Kirchoff's voltage law, the total voltage which appears across them (and the current source) is $V_2 + 1.5V_2 = 2.5V_2$

Since current flows **out** the positive terminal of the current source, we may consider it a negative resistance, of value

$$\frac{-2.5V_2}{V_2/4} = -10\Omega$$

We now reduce the circuit by series-parallel combinations. 8Ω in 12Ω in series add to 20 ohms. It is in parallel with -10Ω so we write $R_{11} = \frac{20 \times (-10)}{20-10} = \frac{-200}{10} = -20\Omega$

This combines in series with 4Ω and 6Ω, yielding $(4+6-20)\Omega = -10\Omega$

The -10Ω is in parallel with 2Ω, so combined resistance is: $\frac{2(-10)}{2-10} = \frac{-20}{-8} = \boxed{2.5\Omega}$

1-45

In the circuit below determine the input resistance across nodes A and B: Given R_1, R_2, k and i_x (all positive numbers).

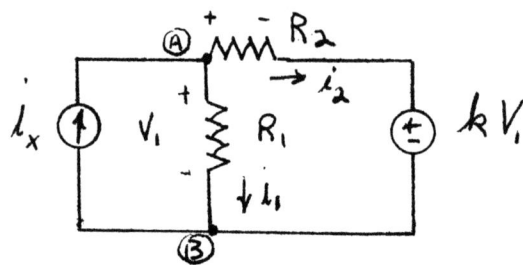

****

From OHM'S Law:

$$R_{AB} = V_{AB}/i_x$$

$$V_{AB} = V_1 = i_1 R_1$$

Determine i_1 using KVL and KCL.

At Node A: KCL yields:

$$i_x = i_1 + i_2 \qquad \qquad 1)$$

KVL around middle and branch on the right, yields:

$$i_1 R_1 - i_2 R_2 - k V_1 = 0 \qquad \qquad 2)$$

From Ohm's Law

$$V_1 = i_1 R_1$$

Thus: $i_1 R_1 - i_2 R_2 - k\, i_1 R_1 = 0$

and

$$i_1 (R_1)(1 - k) - i_2 R_2 = 0 \qquad \qquad 3)$$

Solving simultaneously for i_1:

$$i_1 = i_x / (1 + (R_1/R_2) \cdot (1-K)) \; ; \quad \therefore R_{AB} = R_1 / (1 + (R_1/R_2) \cdot (1-K))$$

1-46

Find v_o/v_s. Assume that the operational amplifiers are ideal.

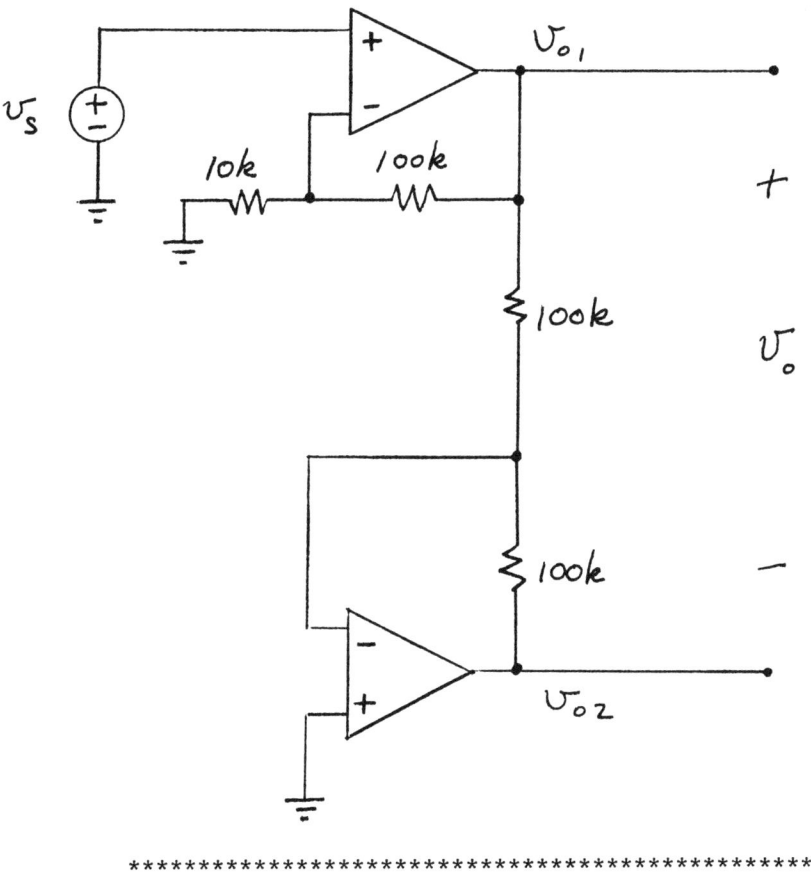

$$v_{o_1} = \frac{10k + 100k}{10k} v_s = \frac{110k}{10k} v_s = 11 v_s$$

$$v_{o_2} = -\frac{100k}{100k} v_{o_1} = -v_{o_1}$$

Now, $v_o = v_{o_1} - v_{o_2} = 11 v_s - (-v_{o_1})$
$= 11 v_s - (-11 v_s)$
$= 22 v_s$

THUS $v_o/v_s = 22$

1-47

When I2 = 10 amperes what is the value of R?

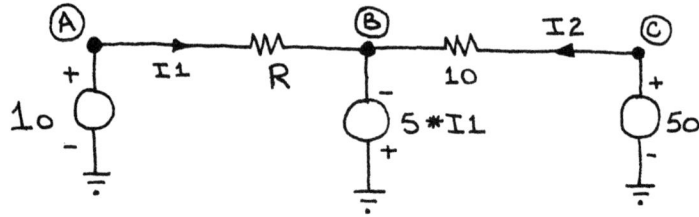

***

Knowing $I2 = 10$, KVL yields $50 = I2*10 - 5*I1$
$$-50 = -5*I1 \quad \text{or} \quad I1 = 10.$$

Using KVL again $10 = I1*R - 5*I1 = I1(R-5) = 10(R-5)$
yielding $R = 6$ ohms.

1-48

Find V in the shown circuit.

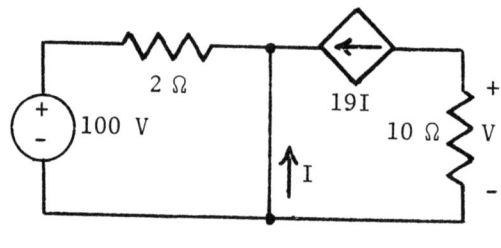

***

By Ohm's law, $V = -10(19I) = -190 I$. So, we need I to determine V. From KCL applied to the top middle node, we see that the current <u>to the left</u> in the 2-Ω resistor is $19I + I = 20I$. We can obtain a numerical value for this current by observing that the short circuit in the middle of the circuit places all the source 100 V across the 2-Ω resistor, thereby producing a resistor current of $100/2 = 50$ A <u>to the right</u>. Consequently,

$$20I = -50 \quad \text{and so} \quad I = -2.5 A$$

Finally, $\quad V = -190 I = -190(-2.5) = 475 \cdot V$

1-49

Determine the voltage drop V_{AO} and the power absorbed by the controlled source.

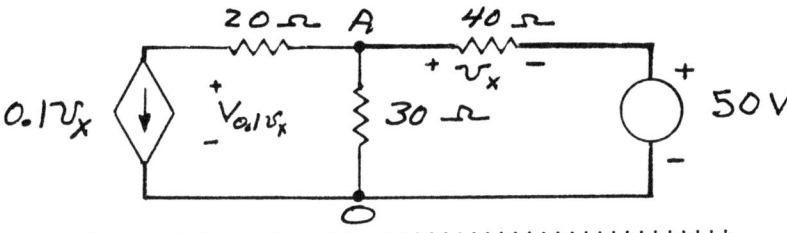

KCL Node A: $+0.1v_x + \frac{1}{30}V_{AO} + \frac{1}{40}(V_{AO} - 50) = 0$

The current in the 20 Ω resistor is that of the controlled current source.

Substituting $v_x = V_{AO} - 50$

$$0.1(V_{AO} - 50) + \frac{1}{30}V_{AO} + \frac{1}{40}(V_{AO} - 50) = 0$$

Solving: $V_{AO} = 39.475$ V

Controlled source current $= 0.1(39.475 - 50)$
$= -1.0525$ A

Writing KVL starting at O going to A

$$-39.475 + 20(-1.0525) + V_{0.1v_x} = 0$$

The voltage drop is $V_{0.1v_x} = 60.525$ V

Controlled source power: $P_{0.1v_x} = (-1.0525)(60.525)$
$= -63.703$ W

1-50

Find the Norton equivalent circuit at terminals A,B.

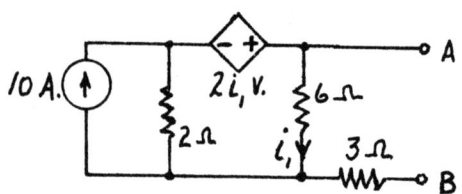

****

TO FIND R_{TH}, USE $R_{TH} = \dfrac{V_{OC}}{I_{SC}}$:

WRITE KVL, LET $V_{OC} = V_{AB}$, IN DIRECTION OF i_1 WE HAVE:

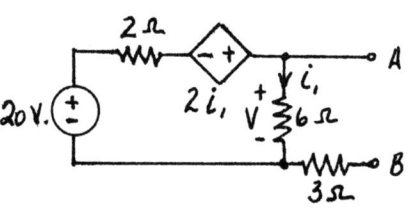

$6i_1 - 20 + 2i_1 - 2i_1 = 0$

$\therefore i_1 = \dfrac{20}{6} = \dfrac{10}{3}$ A.

$\therefore V_{OC} = 6i_1 = 20$ V.

IF WE SHORT THE A-B TERMINALS, $I_{AB} = I_{SC}$, DEFINING V THE VOLTAGE ACROSS THE 6Ω, WE HAVE:

KCL: $\dfrac{V}{6} + \dfrac{V}{3} + \dfrac{V - 2i_1 - 20}{2} = 0$ AND $i_1 = \dfrac{V}{6}$

$V + 2V + 3V - 6i_1 - 60 = 0$

$6V - 6\left(\dfrac{V}{6}\right) = 60$

$5V = 60$

$V = 12$ V. $\therefore I_{SC} = \dfrac{V}{3} = \dfrac{12}{3} = 4$ A.

THEN, $R_{TH} = \dfrac{V_{OC}}{I_{SC}} = \dfrac{20}{4} = 5$ Ω

NORTON CIRCUIT: 4A ↑ 5Ω A, B

2
GENERAL ANALYSIS METHODS

NODAL ANALYSIS

--- 2-1

Determine the voltage drops V_{AO} and V_{BO}.

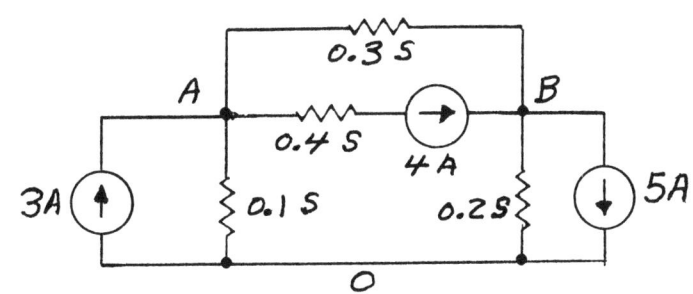

****

$$\begin{pmatrix}\text{KCL} \\ \sum I_{out}=0 \\ \text{node}\end{pmatrix}$$
Node A: $\quad -3 + 0.1 V_{AO} + 4 + 0.3(V_{AO} - V_{BO}) = 0$
Node B: $\quad -4 + 0.3(V_{BO} - V_{AO}) + 0.2 V_{BO} + 5 = 0$

Simplifying:
$$0.4 V_{AO} - 0.3 V_{BO} = -1$$
$$-0.3 V_{AO} + 0.5 V_{BO} = -1$$

Solving:

$$V_{AO} = \frac{\begin{vmatrix} -1 & -0.3 \\ -1 & +0.5 \end{vmatrix}}{\begin{vmatrix} 0.4 & -0.3 \\ -0.3 & +0.5 \end{vmatrix}} = \frac{-0.5 - 0.3}{0.2 - 0.09} = -7.273 V$$

$$V_{BO} = \frac{\begin{vmatrix} 0.4 & -1 \\ -0.3 & -1 \end{vmatrix}}{0.11} = \frac{-0.4 - 0.3}{0.11} = -6.364 V$$

2-2

Use nodal analysis to find the voltage across the 4 ohm resistor.

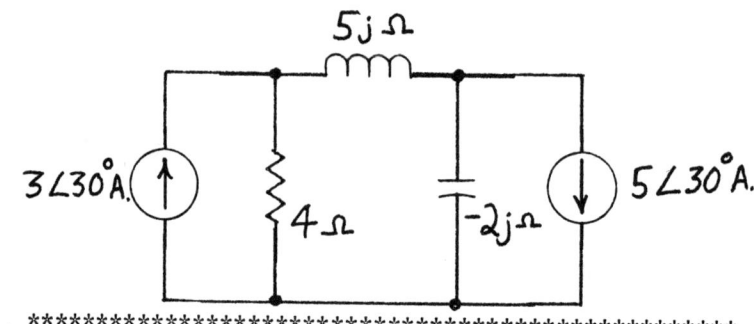

Convert Impedance to Admittance.

Note: S or ℧ (some texts use latter.)

CURRENT INTO NODE 1 = CURRENT OUT OF NODE 1

$$3\angle 30° = \vec{V_1}(.25) + (\vec{V_1} - \vec{V_2})(-.2j)$$

CURRENT INTO NODE 2 = CURRENT OUT OF NODE 2

$$-5\angle 30° = \vec{V_2}(.5j) + (\vec{V_2} - \vec{V_1})(-.2j)$$

SIMPLIFIED:

$$3\angle 30° = \vec{V_1}(.25 - .2j) + \vec{V_2}(.2j)$$

$$-5\angle 30° = \vec{V_1}(.2j) + \vec{V_2}(.3j)$$

IN POLAR FORM:

$$3\angle 30° = \vec{V_1}(.32\angle -38.7°) + \vec{V_2}(.2\angle 90°)$$

$$-5\angle 30° = \vec{V_1}(.2\angle 90°) + \vec{V_2}(.3\angle 90°)$$

$$\vec{V_1} = \frac{\begin{vmatrix} 3\angle 30° & .2\angle 90° \\ -5\angle 30° & .3\angle 90° \end{vmatrix}}{\begin{vmatrix} .32\angle -38.7° & .2\angle 90° \\ .2\angle 90° & .3\angle 90° \end{vmatrix}} = \frac{.9\angle 120° + 1.0\angle 120°}{.096\angle 51.3° - .04\angle 180°}$$

$$= \frac{1.9\angle 120°}{.06 + .075j + .04} = \frac{1.9\angle 120°}{.1 + .075j} = \frac{1.9\angle 120°}{.125\angle 36.8°}$$

$$= \underline{15.2\angle 83.2° \text{ VOLTS}} = \text{VOLTAGE ACROSS THE } 4\Omega \text{ RESISTOR}$$

2-3

Use nodal analysis to find the transfer function V_2/V_s.

[Circuit diagram: V_s source connected through 1kΩ to node V_1, through 1kΩ to node V_2; 1kΩ from V_1 to ground; dependent current source $g_m V_1$; 1kΩ from V_2 to ground. $g_m = \frac{1}{2} \times 10^{-3}$ S (℧)]

****

$$\frac{V_1 - V_s}{1k} + \frac{V_1 - V_2}{1k} + \frac{V_1}{1k} = 0 \rightarrow 3V_1 - V_2 = V_s$$

$$\frac{V_2 - V_1}{1k} + \frac{V_2}{1k} + g_m V_1 = 0 \rightarrow -\frac{1}{2}V_1 + 2V_2 = 0$$

$$V_2 = \frac{\begin{vmatrix} 3 & V_s \\ -\frac{1}{2} & 0 \end{vmatrix}}{\begin{vmatrix} 3 & -1 \\ -\frac{1}{2} & 2 \end{vmatrix}} = \frac{+\frac{1}{2}V_s}{6 - \frac{1}{2}} = \frac{V_s}{11}$$

$$\boxed{\frac{V_2}{V_s} = \frac{1}{11}}$$

2-4

Determine i by the nodal method.

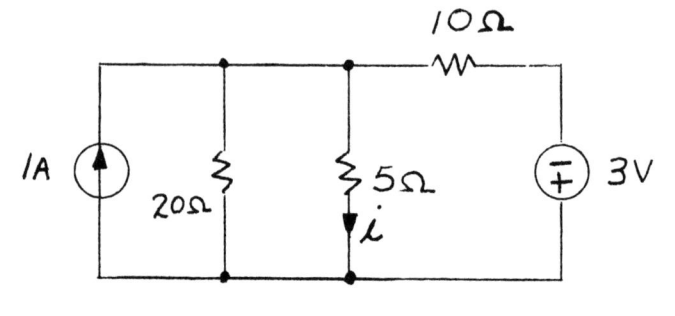

***

Letting V be the voltage across the 5Ω resistor, we have that, by the nodal method,

$\sum I_{out} = 0$ top node

$$-1 + \frac{V}{20} + \frac{V}{5} + \frac{V-(-3)}{10} = 0$$

Solving this equation for V results in

$$V = 2 \text{ V}.$$

Then,

$$i = \frac{V}{5} = \frac{2}{5} = 0.4 \text{ A}.$$

Note that only one nodal equation was required to solve this problem.

2-5

Calculate the total power absorbed by the resistors.

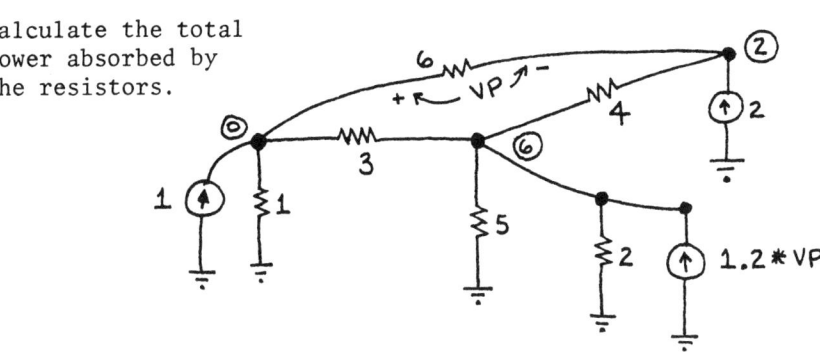

Using node equation for the reference indicated

$$\begin{bmatrix} 1.5 & -.333 & -.167 & 0 \\ -.333 & 1.28 & -.25 & -1.2 \\ -.167 & -.25 & .4167 & 0 \\ -1 & 0 & 1 & 1 \end{bmatrix} * \begin{bmatrix} V0 \\ V6 \\ V2 \\ VP \end{bmatrix} = \begin{bmatrix} 1 \\ 0 \\ 2 \\ 0 \end{bmatrix} \quad \text{Solving} \begin{bmatrix} V0 \\ V6 \\ V2 \\ VP \end{bmatrix} = \begin{bmatrix} .631 \\ -2.07 \\ 3.81 \\ -3.18 \end{bmatrix}$$

Total pwr Absorbed = 16.15 watts
$= (.63)1 + (1.2)(-2.07)(-3.18) + 2(3.81) = 16.15$

2-6

Calculate the value of the current source required such that node-pair voltage $V(y,w) = 3.25$.

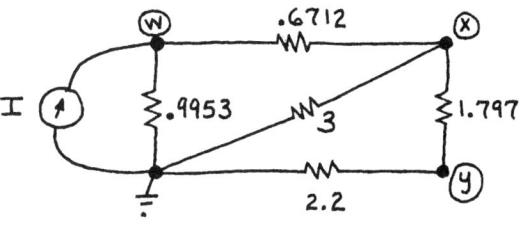

Write the node equations using the reference node n model.

$$\begin{bmatrix} 2.49 & -1.49 & 0 & -1 \\ -1.49 & 2.38 & -0.557 & 0 \\ 0 & -.557 & 1.01 & 0 \\ 1 & 0 & -1 & 0 \end{bmatrix} \begin{bmatrix} Vw \\ Vx \\ Vy \\ I \end{bmatrix} = \begin{bmatrix} 0 \\ 0 \\ 0 \\ 3.25 \end{bmatrix}$$

Solving for $I = 7.66$

2-7

For the circuit below find v using node analysis.

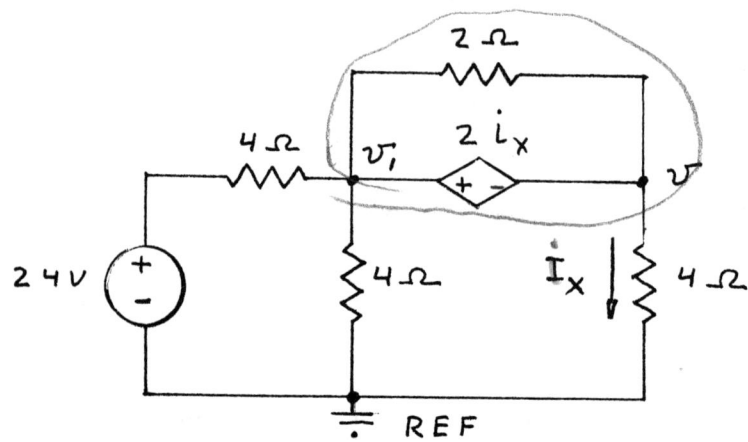

****

SUPERNODE $v_1 - v$: $\quad \dfrac{v_1 - 24}{4} + \dfrac{v_1}{4} + \dfrac{v}{4} = 0 \quad ①$

(NOTE: CURRENT THROUGH 2Ω RESISTOR BOTH ENTERS AND LEAVES THIS NODE.)

$$\text{KVL} \quad v_1 - v = 2i_x$$

$$\text{BUT } i_x = \dfrac{v}{4}$$

$$\text{SO} \quad v_1 - v = 2\dfrac{v}{4}$$

$$\text{OR} \quad v_1 = 1.5\, v \quad\quad ②$$

SOLVING ① & ② FOR v

$$v = 6 \text{ VOLTS} \quad\quad \text{ANS}$$

2-8

Determine the value of the current i_x using the nodal analysis method.

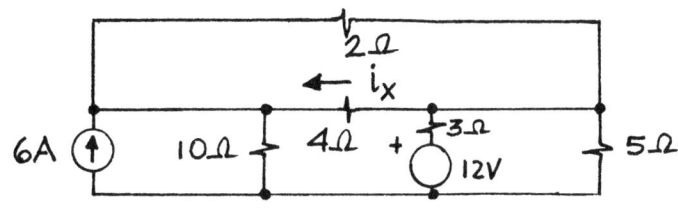

***

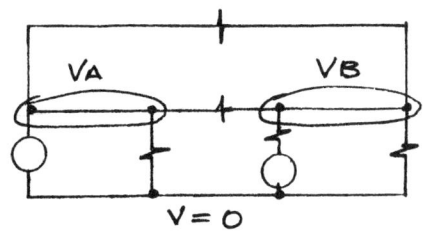

SELECTED NODES

KCL EQS AT NODES A AND B

AT A: $\quad \dfrac{V_A}{10} + \dfrac{V_A - V_B}{2} + \dfrac{V_A - V_B}{4} = 6$

AT B: $\quad \dfrac{V_B}{5} + \dfrac{V_B - 12}{3} + \dfrac{V_B - V_A}{2} + \dfrac{V_B - V_A}{4} = 0$

A: $\quad 0.85\, V_A - 0.75\, V_B = 6$
B: $\quad -0.75\, V_A + 1.28\, V_B = 4$

$$V_A = \dfrac{\begin{vmatrix} 6 & -0.75 \\ 4 & 1.28 \end{vmatrix}}{\begin{vmatrix} 0.85 & -0.75 \\ -0.75 & 1.28 \end{vmatrix}} = \dfrac{10.68}{0.53} = 20.15\,V \quad ; \quad V_B = 14.91\,V$$

$$i_x = \dfrac{V_B - V_A}{4} = \dfrac{14.91 - 20.15}{4} = -1.31\,A$$

2-9

Given the following circuit, find the node voltages and the current through the 2 Ω resistor.

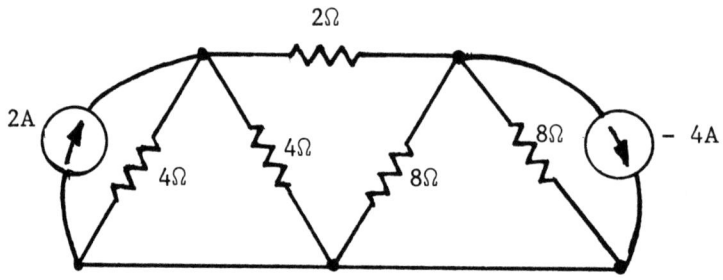

****

The Circuit Can be Simplified to the form:

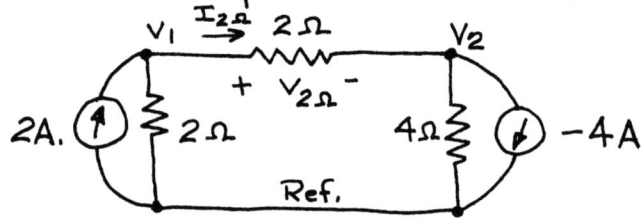

Kirchhoff's Current Law Gives:

Node #1: $2 = \frac{1}{2}V_1 + \frac{1}{2}(V_1 - V_2)$
Node #2: $-(-4) = \frac{1}{4}V_2 + \frac{1}{2}(V_2 - V_1)$

These Equations Can be Rewritten As

$$2 = V_1 - \frac{1}{2}V_2$$
$$4 = -\frac{1}{2}V_1 + \frac{3}{4}V_2$$

By Cramer's Rule

$$V_1 = \frac{\begin{vmatrix} 2 & -\frac{1}{2} \\ 4 & \frac{3}{4} \end{vmatrix}}{\begin{vmatrix} 1 & -\frac{1}{2} \\ -\frac{1}{2} & \frac{3}{4} \end{vmatrix}} = \frac{\frac{3}{2} + 2}{\frac{3}{4} - \frac{1}{4}} = \frac{\frac{7}{2}}{\frac{1}{2}} = 7V.$$

$$V_2 = \frac{\begin{vmatrix} 1 & 2 \\ -1/2 & 4 \end{vmatrix}}{1/2} = \frac{4+1}{1/2} = 10 \text{ V}.$$

$V_{2\Omega}$ Is Just The Voltage Difference Between Nodes.

$$V_{2\Omega} = V_1 - V_2 = 7 - 10 = -3 \text{ V}.$$

$$I_{2\Omega} = V_{2\Omega}/2 = -3/2 \text{ A}.$$

2-10

The node equations are shown. Provide a network schematic whose behaviors satisfy these equations and label each component value.

$$\begin{bmatrix} 0.5 & 0 & 0 & 0 & -1 & 0 \\ 0 & 1.333 & -1 & -.333 & 1 & 0 \\ 0 & -1 & 1 & 0 & 0 & 0 \\ 0 & -.333 & 0 & .333 & 0 & -1 \\ 1 & -1 & 0 & 0 & 0 & 0 \\ -5 & 0 & 0 & 1 & 0 & 0 \end{bmatrix} \cdot \begin{bmatrix} V_1 \\ V_X \\ V_2 \\ V_3 \\ I_Y \\ I_X \end{bmatrix} = \begin{bmatrix} 0 \\ 0 \\ 2 \\ 0 \\ 4 \\ 0 \end{bmatrix}$$

****

Schematic

The node equations using the reference node 0 model show five nodes (0 which is reference, 1, X, 2, and 3). Note the presence of one CIS at (2,0), one CVS of $V(1,X) = 4$ and a VCVS at (3,0). The conductance at node-pair (X,3) is 0.333 mhos yielding the labeled resistance of 3 ohms. You can now complete the labeling of the component values.

2-11

Using the method of nodal analysis, find v_1, v_2, v_3 for this circuit.

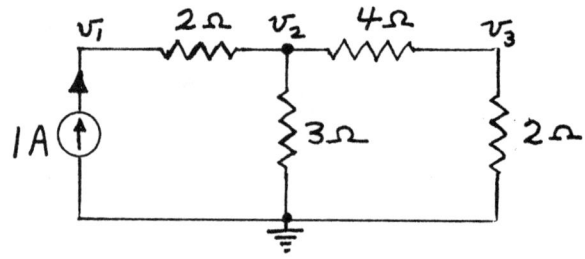

****

KCL at node v_1: $\quad \dfrac{v_1 - v_2}{2} = 1 \quad$ (eq. 1)

KCL at node v_2: $\quad 1 + \dfrac{v_3 - v_2}{4} = \dfrac{v_2}{3} \quad$ (eq. 2)

KCL at node v_3: $\quad \dfrac{v_2 - v_3}{4} = \dfrac{v_3}{2} \quad$ (eq. 3)

Solving by elimination:

From (eq. 3) $\quad v_2 = 3 v_3$

Substituting into (eq. 2):

$1 + \dfrac{v_3}{4} = \dfrac{3 v_3}{4} + \dfrac{3 v_3}{3} \quad \therefore \; \underline{v_3 = \tfrac{2}{3} V} \text{ and } \underline{v_2 = 2V}$

From (eq. 1) $\quad \underline{v_1 = 4V}$

As a check:

(eq. 2) $\quad 1 + \dfrac{\tfrac{2}{3} - 2}{4} = \dfrac{2}{3} \quad \checkmark$

(eq. 3) $\quad \dfrac{2 - \tfrac{2}{3}}{4} = \dfrac{\tfrac{2}{3}}{2} \quad \checkmark$

Nodal Analysis / 51

2-12

Write node equations and auxiliary equations (if needed) so that all the node voltages of the networks shown can be determined in terms of the constants of the network (Rs, Ls, Cs, Ks of dependent sources) and independent sources. Do not solve the equations.

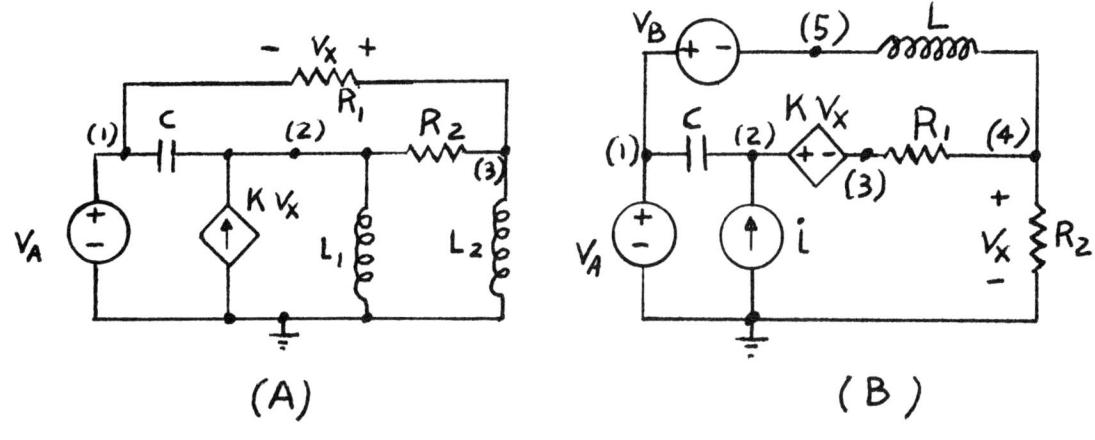

(A) (B)

***

NETWORK A

NODE 2: $C\dfrac{d(V_2 - V_A)}{dt} - KV_x + \dfrac{1}{L_1}\int_0^t V_2\, dt + \dfrac{V_2 - V_3}{R_2} = 0$

NODE 3: $\dfrac{1}{L_2}\int_0^t V_3\, dt + \dfrac{V_3 - V_2}{R_2} + \dfrac{V_3 - V_A}{R_1} = 0$

AUXILIARY EQUATION: $V_A + V_x - V_3 = 0$
(USE TO ELIMINATE V_x FROM NODE EQUATION).

NETWORK B

NODE 2: $C\dfrac{d(V_2 - V_A)}{dt} - i + \dfrac{V_2 - KV_x - V_x}{R_1} = 0$

NODE 4: $\dfrac{V_x}{R_2} + \dfrac{V_x + KV_x - V_2}{R_1} + \dfrac{1}{L}\int_0^t (V_x + V_B - V_A)\, dt = 0$

AN AUXILIARY EQUATION IS NOT NEEDED SINCE V_x IS A NODE VOLTAGE.

2-13

In this circuit determine v_1 and v_2 using the node voltage equations.

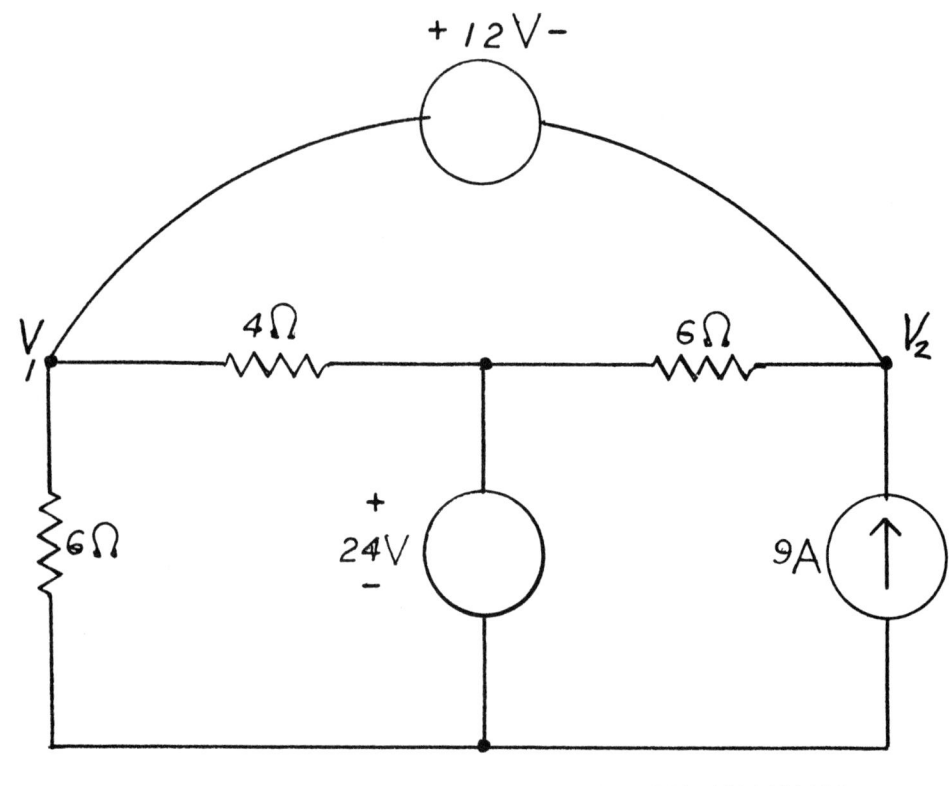

Write Kirchoff's current law at the super-node consisting of V_1 and V_2 and 12V source. The voltage at the center node is known to be 24 Volts. $V_2 = V_1 - 12$

$$\frac{V_1}{6} + \frac{V_1-24}{4} + \frac{V_2-24}{6} = 9 = \frac{V_1}{6} + \frac{V_1}{4} + \frac{V_1}{6} - \frac{24}{4} - \frac{36}{6}$$

Multiply through by 12

$2V_1 + 3V_1 + 2V_1 - 72 - 72 = 108$

$7V_1 - 144 = 108$

$7V_1 = 252$

$V_1 = \underline{36 \text{ volts}}$

$V_2 = V_1 - 12 = \underline{24 \text{ volts}}$

2-14

Using nodal analysis, determine the node c voltage.

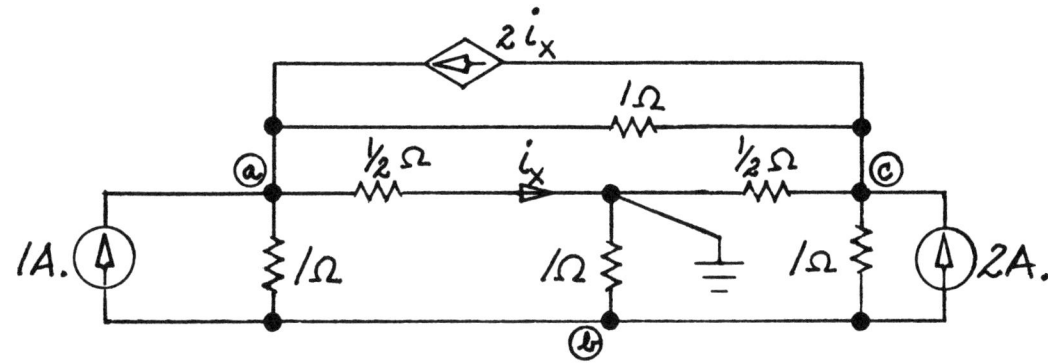

****

$\underline{KCL\,@\,a}$: $-1 + \dfrac{v_a - v_b}{1} + \dfrac{v_a}{1/2} + \dfrac{v_a - v_c}{1} - 2i_x = 0 \;\; \& \;\; i_x = \dfrac{v_a}{1/2} = 2v_a$

$\Rightarrow -v_b - v_c = 1$

$\underline{KCL\,@\,b}$: $1 + \dfrac{v_b - v_a}{1} + \dfrac{v_b}{1} + \dfrac{v_b - v_c}{1} + 2 = 0$

$\Rightarrow -v_a + 3v_b - v_c = -3$

$\underline{KCL\,@\,c}$: $\dfrac{v_c}{1/2} + \dfrac{v_c - v_a}{1} + 2i_x + \dfrac{v_c - v_b}{1} - 2 = 0 \;\; \& \;\; i_x = \dfrac{v_a}{1/2} = 2v_a$

$\Rightarrow 3v_a - v_b + 4v_c = 2$

Solving the three eqns. simultaneously yields $\underline{v_c = -1/7 \, V.}$

2-15

In the following circuit, all resistances are 4 kΩ. Find v_o.

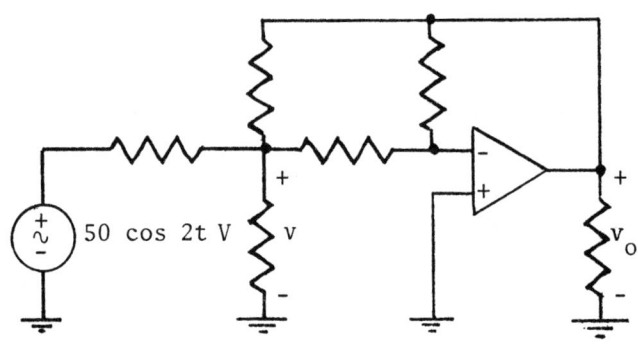

***

Since for this circuit, voltage is the input and voltage is the desired output, we can scale all resistances to a convenient value, say 1 Ω, and obtain the correct v_o. From applying nodal analysis to the v node we obtain

$$4v - v_o = 50 \cos 2t$$

in which we used a conductance of 1 S for each resistor. We also used the fact that the inverting input terminal is at a virtual ground. Because this equation has two unknowns, we need another equation. Applying nodal analysis at the inverting input terminal by summing currents into this node, we obtain

$$v + v_o = 0$$

from which we determine that $v = -v_o$. Substituting $-v_o$ for v in the v nodal equation, we obtain

$$4(-v_o) - v_o = 50 \cos 2t$$

And so

$$v_o = -10 \cos 2t \text{ V}$$

2-16

Find the current I_L that flows through the resistor R_L using Nodal Analysis.

[Circuit: $E_1 = 42v$ with $R_1 = 3\,\Omega$, $R_L = 6\,\Omega$ (with I_L flowing down), $R_2 = 12\,\Omega$, $E_2 = 21v$]

***

Establish the Nodal Equation for Nodal Voltage V_A using Kirchhoff's Current Law.

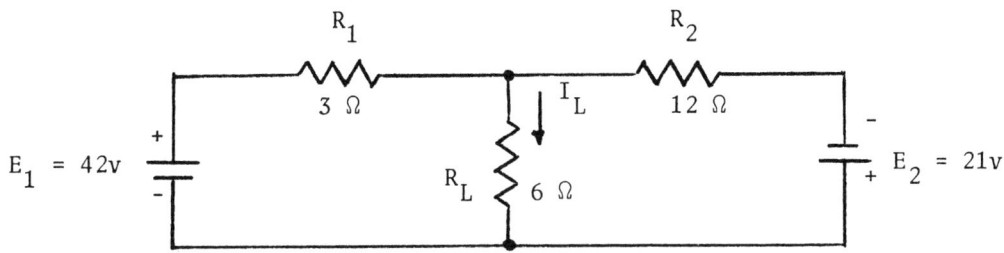

$$I_1 - I_L - I_2 = 0$$

by substitution

$$\frac{E_1 - V_A}{R_1} - \frac{V_A}{R_L} - \left(\frac{V_A - (-E_2)}{R_2}\right) = 0$$

$$\frac{E_1}{R_1} - \frac{V_A}{R_1} - \frac{V_A}{R_L} - \frac{V_A}{R_2} - \frac{E_2}{R_2} = 0$$

$$\left(\frac{1}{R_1} + \frac{1}{R_L} + \frac{1}{R_2}\right) V_A = \frac{E_1}{R_1} - \frac{E_2}{R_2}$$

$$V_A = \frac{\frac{E_1}{R_1} - \frac{E_2}{R_2}}{\frac{1}{R_1} + \frac{1}{R_L} + \frac{1}{R_2}} = \frac{\frac{42V}{3\Omega} - \frac{21V}{12\Omega}}{\frac{1}{3\Omega} + \frac{1}{6\Omega} + \frac{1}{12\Omega}} = 21V$$

$$I_L = \frac{V_A}{R_L} = \frac{21V}{6\Omega} = 3.5a$$

2-17

For the circuit below, write the nodal equations which will allow you to solve for the four node voltages and then find v_4. All element values are in Siemens.

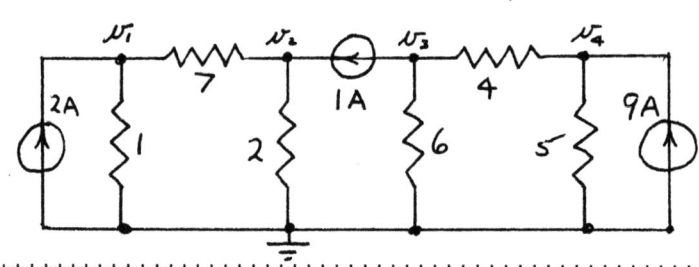

***

$$8v_1 - 7v_2 = 2$$

$$-7v_1 + 9v_2 = 1$$

$$10v_3 - 4v_4 = -1$$

$$-4v_3 + 9v_4 = 9$$

$$v_4 = \frac{\begin{vmatrix} 10 & -1 \\ -4 & 9 \end{vmatrix}}{\begin{vmatrix} 10 & -4 \\ -4 & 9 \end{vmatrix}} = \frac{90-4}{90-16}$$

$$v_4 = \frac{86}{74} = 1.16 \text{ V}$$

2-18

Write the nodal equations required for determining the shown node voltages in the following circuit.

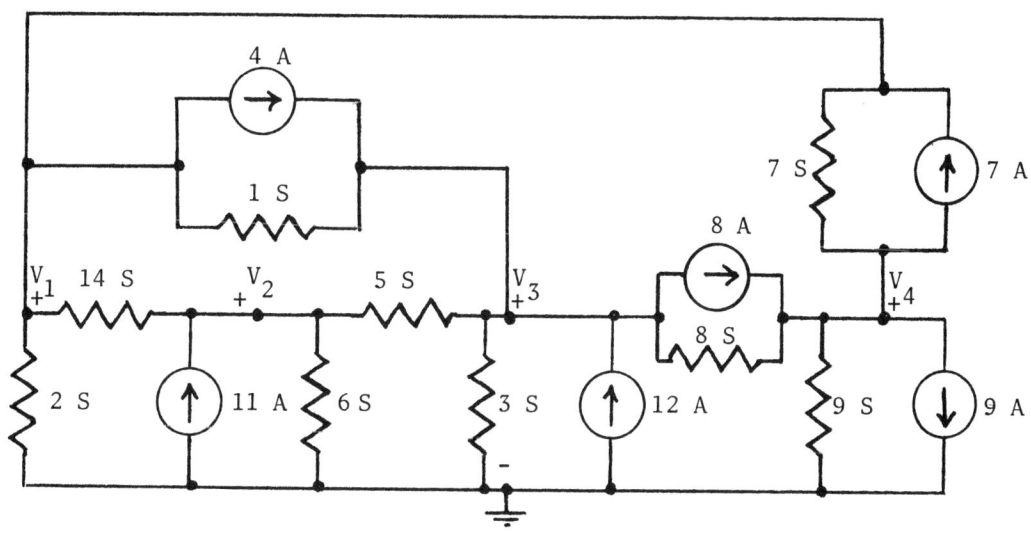

***

The sum of the currents from current sources into the non grounded nodes are $7-4 = 3$ A for the V_1 node, 11 A into the V_2 node, $4+12-8 = 8$ A into the V_3 node, and $8-7-9 = -8$ A into the V_4 node. The nodal self-admittances are

$$2+14+1+7 = 24 \text{ S for the } V_1 \text{ node}$$
$$14+6+5 = 25 \text{ S for the } V_2 \text{ node}$$
$$1+5+3+8 = 17 \text{ S for the } V_3 \text{ node}$$
$$8+9+7 = 24 \text{ S for the } V_4 \text{ node}$$

The mutual admittances are

14 S for the V_1, V_2 nodes; 1 S for the V_1, V_3 nodes;
7 S for the V_1, V_4 nodes; 5 S for the V_2, V_3 nodes;
0 S for the V_2, V_4 nodes; 8 S for the V_3, V_4 nodes;

So, the nodal equations are

$$24V_1 - 14V_2 - V_3 - 7V_4 = 3$$
$$-14V_1 + 25V_2 - 5V_3 \qquad\quad = 11$$
$$-V_1 - 5V_2 + 17V_3 - 8V_4 = 8$$
$$-7V_1 \qquad\quad - 8V_3 + 24V_4 = -8$$

2-19

Using nodal equations, determine the voltage across resistor R_5.

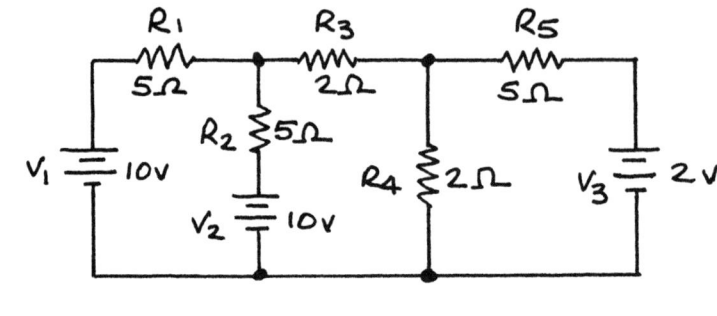

****

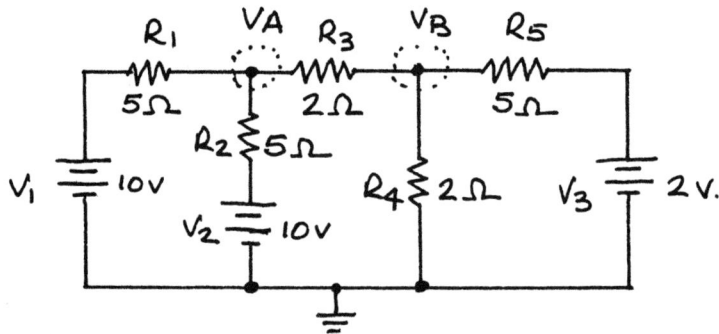

$$V_{R_5} = V_B - V_3$$

Writing Kirchhoff's current law at the two nodes, A and B, gives

$$\frac{V_A - V_1}{R_1} + \frac{V_A - V_2}{R_2} + \frac{V_A - V_B}{R_3} = 0$$

$$\frac{V_B - V_A}{R_3} + \frac{V_B}{R_4} + \frac{V_B - V_3}{R_5} = 0$$

Combining terms

$$V_A \left(\frac{1}{R_1} + \frac{1}{R_2} + \frac{1}{R_3}\right) - V_B \left(\frac{1}{R_3}\right) = \frac{V_1}{R_1} + \frac{V_2}{R_2}$$

$$- V_A \left(\frac{1}{R_3}\right) + V_B \left(\frac{1}{R_3} + \frac{1}{R_4} + \frac{1}{R_5}\right) = \frac{V_3}{R_5}$$

Substituting values and solving for VB:

$$V_A(0.9) - V_B(0.5) = 4$$
$$-V_A(0.5) + V_B(1.2) = 0.4$$

$$V_B = \frac{\begin{vmatrix} 0.9 & 4 \\ -0.5 & 0.4 \end{vmatrix}}{\begin{vmatrix} 0.9 & -0.5 \\ -0.5 & 1.2 \end{vmatrix}} = \frac{2.36}{0.83} = 2.84 \text{ volts}$$

$$V_{R_5} = V_B - V_3 = 2.84 - 2 = 0.84 \text{ volts}$$

$$\boxed{V_{R_5} = 0.84 \text{ volts}}$$

MESH ANALYSIS

2-20

In the network below, find I_{R_5} by use of mesh equations.

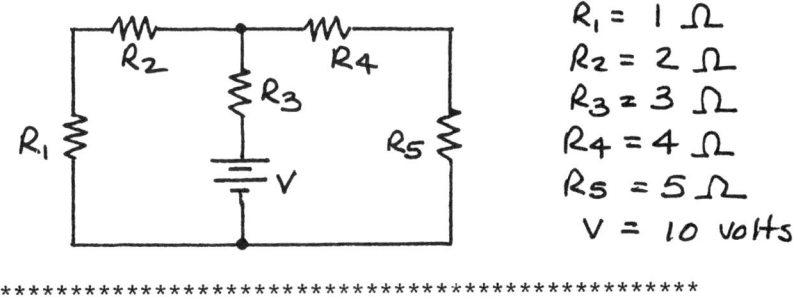

$R_1 = 1\,\Omega$
$R_2 = 2\,\Omega$
$R_3 = 3\,\Omega$
$R_4 = 4\,\Omega$
$R_5 = 5\,\Omega$
$V = 10 \text{ volts}$

****

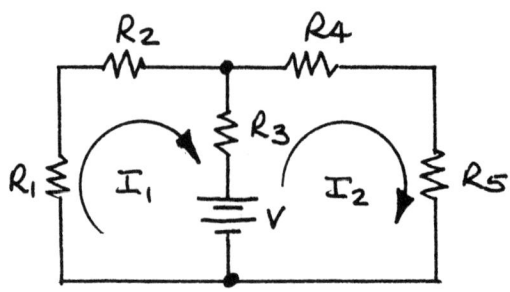

$IR_5 = I_2$

Writing Kirchhoff's voltage law for each mesh

$$-V = I_1(R_1 + R_2 + R_3) - I_2(R_3)$$
$$V = -I_1(R_3) + I_2(R_3 + R_4 + R_5)$$

Substituting values

$$-10 = I_1(6) - I_2(3)$$
$$10 = -I_1(3) + I_2(12)$$

Solving for I_2

$$I_2 = \frac{\begin{vmatrix} 6 & -10 \\ -3 & 10 \end{vmatrix}}{\begin{vmatrix} 6 & -3 \\ -3 & 12 \end{vmatrix}} = \frac{30}{63} = 0.476 \text{ Amps}$$

$$\boxed{IR_5 = 476 \text{ mA}}$$

2-21

Solve for the voltage v by:
a) Loop (mesh) equations.
b) Nodal analysis.

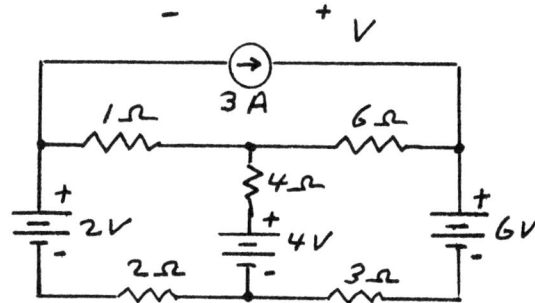

****

a)
$$\begin{cases} 2-4 = 7I_1 - 4I_2 - I_3 \\ 4-6 = -4I_1 + 13I_2 - 6I_3 \\ I_3 = 3 \end{cases}$$

$$\begin{cases} 7I_1 - 4I_2 = 1 \\ -4I_1 + 13I_2 = 16 \end{cases} \rightarrow I_1 = 77/75, \ I_2 = 116/75$$

$$V = 6(I_3 - I_2) + 1(I_3 - I_1) = 802/75 = \boxed{10.69}$$

b)
$$V_2 = V_1 + 2$$
$$V_6 = V_5 + 6$$
$$V_3 = 4$$

$$\frac{V_2 - V_4}{1} + \frac{V_2 - 2}{2} = -3$$

$$\frac{V_4 - V_2}{1} + \frac{V_4 - 4}{4} + \frac{V_4 - V_6}{6} = 0$$

$$\frac{V_6 - V_4}{6} + \frac{V_6 - 6}{3} = 3 \longrightarrow V_2 = -4/75, \ V_6 = \frac{798}{75}$$

$$V = V_6 - V_4 = \frac{798}{75} - \left(-\frac{4}{75}\right) = \frac{802}{75} = \boxed{10.69}$$

2-22

Write loop equations and auxiliary equations (if needed) so that all the loop currents of the networks shown can be determined in terms of the constants of the network (Rs, Ls, Cs, Ks of dependent sources) and independent sources. Do not solve the equations.

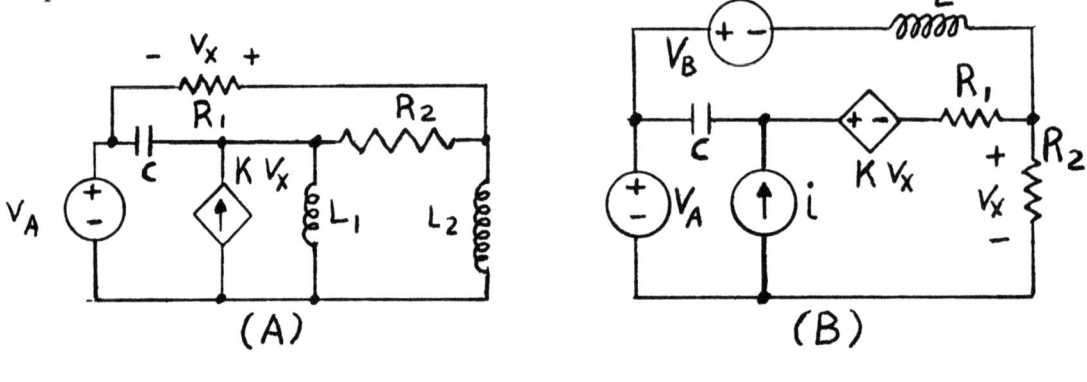

(A)   (B)

****

THE LOOPS CAN BE CHOSEN AS SHOWN:
NETWORK A

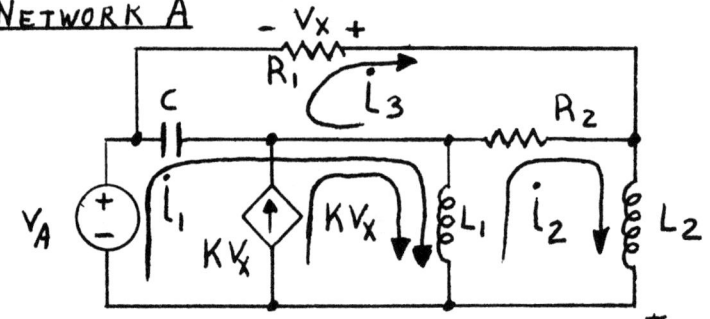

Loop 1: $V_A = \dfrac{1}{C}\int_0^t i_1\,dt + L_1\dfrac{di_1}{dt} - \dfrac{1}{C}\int_0^t i_3\,dt + L_1 K\dfrac{dV_x}{dt} - L_1\dfrac{di_2}{dt}$

Loop 2: $0 = (L_1 + L_2)\dfrac{di_2}{dt} + R_2 i_2 - L_1\dfrac{di_1}{dt} - L_1 K\dfrac{dV_x}{dt} - R_2 i_3$

Loop 3: $0 = i_3(R_1 + R_2) + \dfrac{1}{C}\int_0^t i_3\,dt - \dfrac{1}{C}\int_0^t i_1\,dt - R_2 i_2$

AUXILIARY EQUATION: $V_x = -i_3 R_1$

NETWORK B

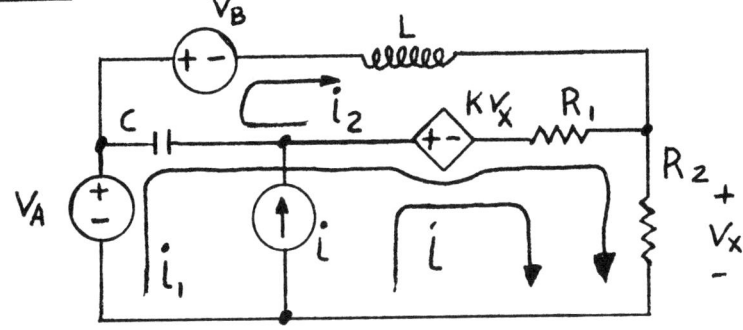

Loop 1: $V_A = \frac{1}{C}\int_0^t i_1\,dt + KV_x + i_1(R_1+R_2) - \frac{1}{C}\int_0^t i_2\,dt$
$\qquad\qquad - i_2 R_1 + i R_2$

Loop 2: $-V_B = L\frac{di_2}{dt} + i_2 R_1 - KV_x + \frac{1}{C}\int_0^t i_2\,dt$
$\qquad\qquad - \frac{1}{C}\int_0^t i_1\,dt - i_1 R_1 - i R_1$

AUXILIARY EQUATION: $\quad V_x = (i_1 + i)R_2$

2-23

Find the currents I_1, I_2, and I_3 shown on the diagram.

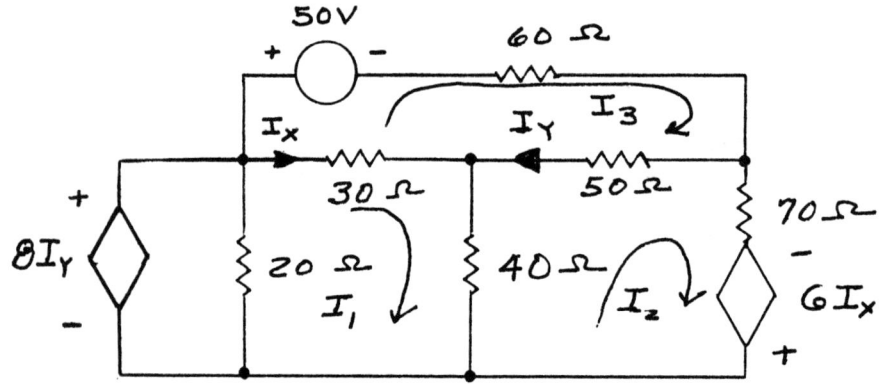

$\left(\sum_{\text{loop}} V_{\text{drops}} = 0\right)$ Note: Can remove 20Ω resistor since in parallel with a voltage source.

KVL Loop 1: $-8I_Y + 30(I_1 - I_3) + 40(I_1 - I_2) = 0$

Loop 2: $40(I_2 - I_1) + 50(I_2 - I_3) + 70 I_2 - 6I_x = 0$

Loop 3: $+50 + 60 I_3 + 50(I_3 - I_2) + 30(I_3 - I_1) = 0$

Substitute: $I_x = I_1 - I_3$ and $I_Y = I_3 - I_2$

Simplify:
$$70 I_1 - 32 I_2 - 38 I_3 = 0$$
$$-46 I_1 + 160 I_2 - 44 I_3 = 0$$
$$-30 I_1 - 50 I_2 + 140 I_3 = -50$$

Solving: $I_1 = -0.4180$; $I_2 = -0.2695$; $I_3 = -0.5430$

2-24

Find the current I_L that flows through the resistor R_L using Mesh Analysis.

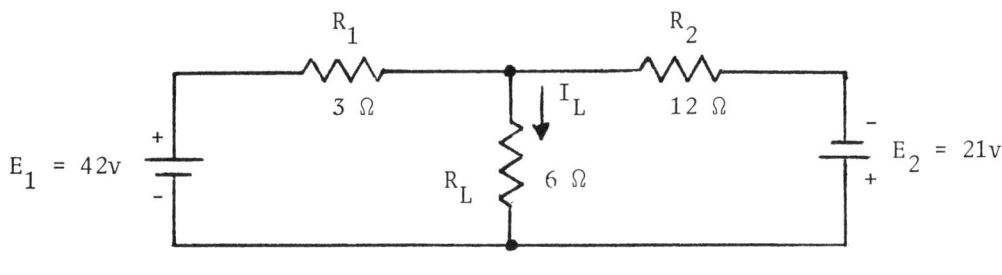

****

Establish Loop Equations using Kirchhoff's Voltage Rule

Loop 1:
$-R_1 I_1 - R_L I_1 + R_L I_2 + E_1 = 0$
$-(R_1 + R_L) I_1 + R_L I_2 = -E_1$
$-(3\Omega + 6\Omega) I_1 + (6\Omega) I_2 = -42v$

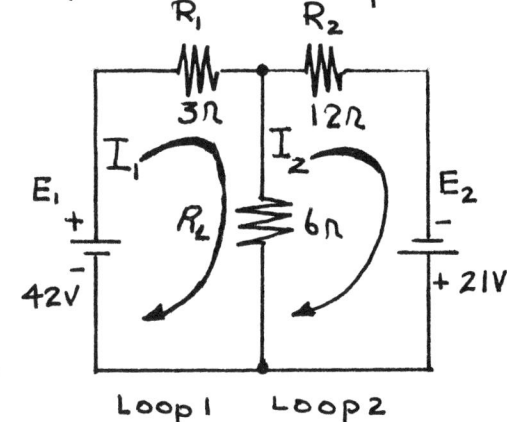

Loop 2:
$-R_2 I_2 + E_2 - R_L I_2 + R_L I_1 = 0$
$+R_L I_1 - (R_2 + R_L) I_2 = -E_2$
$+(6\Omega) - (12\Omega + 6\Omega) I_2 = -21v$

From Loop 1: $(-9\Omega) I_1 + (6\Omega) I_2 = -42v$ ⎫ System of
From Loop 2: $(6\Omega) I_1 - (18\Omega) I_2 = -21v$ ⎬ Simultaneous
                                                    ⎭ Equations

$$I_1 = \frac{\begin{vmatrix} -42 & 6 \\ -21 & -18 \end{vmatrix}}{\begin{vmatrix} -9 & 6 \\ 6 & -18 \end{vmatrix}} = \frac{(-42)(-18) - (-21)(6)}{(-9)(-18) - (6)(6)} = 7a$$

Solve for I_1 and I_2 by using Cramer's Rule
(the use of determinants to find the solution)

$$I_2 = \frac{\begin{vmatrix} -9 & -42 \\ 6 & -21 \end{vmatrix}}{\begin{vmatrix} -9 & 6 \\ 6 & -18 \end{vmatrix}} = \frac{(-9)(-21) - (6)(-42)}{(-9)(-18) - (6)(6)} = 3.5a$$

I_L is the algebraic sum of the mesh currents I_1 and I_2 that flows through R_L

$$I_L = I_1(\downarrow) - I_2(\uparrow) = 7a - 3.5a = 3.5a(\downarrow)$$

2-25

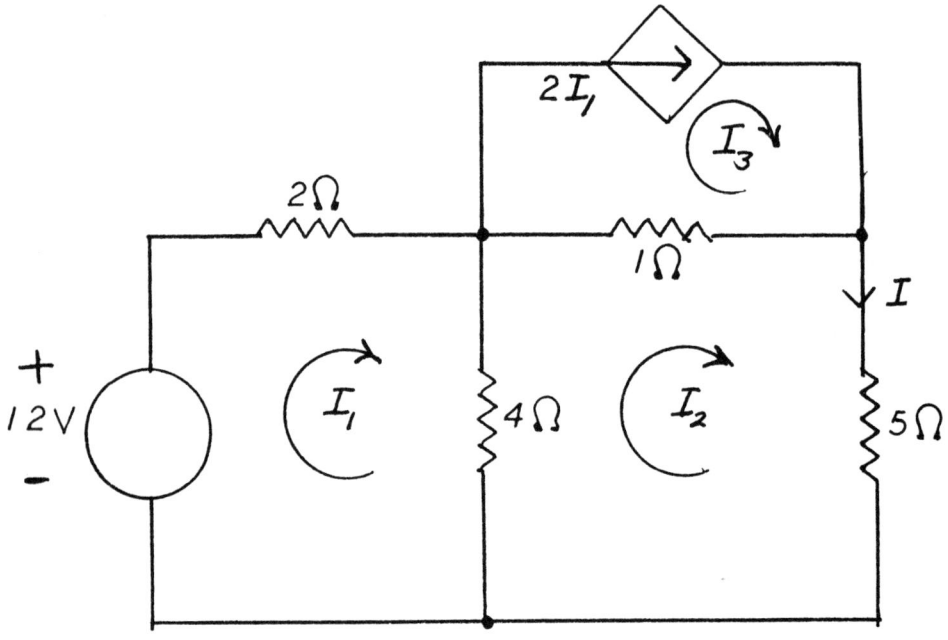

Find I in the circuit above using mesh current analysis and the current variables shown.

***

Around left mesh.

$(4+2)I_1 - 4I_2 = 12$ ①

Around right mesh

$-4I_1 + (4+1+5)I_2 - I_3 = 0$ ②

but $I_3 = 2I_1$

$-6I_1 + 10I_2 = 0$ ②

$6I_1 - 4I_2 = 12$ ① Add

$6I_2 = 12$

$I = I_2 = \underline{2 \text{ Amperes}}$

2-26

Given the following circuit, use mesh analysis to find the mesh currents.

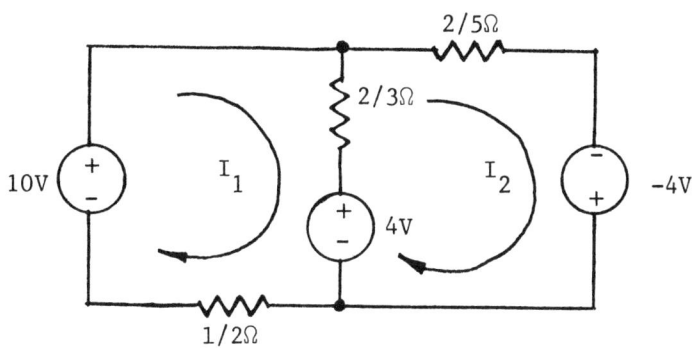

****

For I_1 And I_2 Given, Kirchhoff's Voltage Law Yields

$$-10 + \tfrac{2}{3}(I_1 - I_2) + 4 + \tfrac{1}{2} I_1 = 0$$

$$-4 + \tfrac{2}{3}(I_2 - I_1) + \tfrac{2}{5} I_2 - (-4) = 0$$

If The Equations Are Simplified, Then

$$-(-10+4) = 6 = (\tfrac{2}{3} + \tfrac{1}{2}) I_1 - \tfrac{2}{3} I_2$$

$$-(-4-(-4)) = 0 = -\tfrac{2}{3} I_1 + (\tfrac{2}{5} + \tfrac{2}{3}) I_2$$

Or

$$6 = \tfrac{7}{6} I_1 - \tfrac{2}{3} I_2$$

$$0 = -\tfrac{2}{3} I_1 + \tfrac{16}{15} I_2$$

Now.

$$I_1 = \frac{\begin{vmatrix} 6 & -2/3 \\ 0 & 16/15 \end{vmatrix}}{7/6 \cdot 16/15 - (-2/3)(-2/3)} = \frac{6 \cdot 16/15}{\tfrac{7}{6} \cdot \tfrac{16}{15} - \tfrac{4}{9}} = \frac{6.4}{0.8} = 8A.$$

$$I_2 = \frac{\begin{vmatrix} 7/6 & 6 \\ -2/3 & 0 \end{vmatrix}}{0.8} = \frac{-6(-2/3)}{0.8} = \frac{4}{0.8} = 5A.$$

Check:

$$\Sigma V_s' = -10 + \tfrac{2}{3}(8-5) + 4 + \tfrac{1}{2} \cdot 8 = -10 + 2 + 4 + 4 \equiv 0$$

2-27

By first solving for the mesh currents find v_1 and v_2 in the circuit below, all resistance values are in ohms.

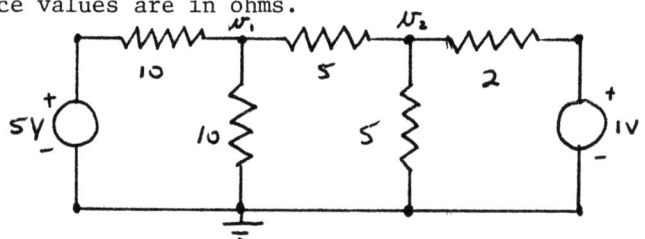

****

$$20 i_1 - 10 i_2 = 5$$

$$-10 i_1 + 20 i_2 - 5 i_3 = 0$$

$$-5 i_2 + 7 i_3 = -1$$

$$i_1 = \frac{\begin{vmatrix} 5 & -10 & 0 \\ 0 & 20 & -5 \\ -1 & -5 & 7 \end{vmatrix}}{\begin{vmatrix} 20 & -10 & 0 \\ -10 & 20 & -5 \\ 0 & -5 & 7 \end{vmatrix}} = \frac{700 - 50 - 125}{2800 - 500 - 700}$$

$$i_1 = \frac{525}{1600} = \frac{21}{64} \text{ A}$$

$$i_2 = \frac{\begin{vmatrix} 20 & 5 & 0 \\ -10 & 0 & -5 \\ 0 & -1 & 7 \end{vmatrix}}{1600} = \frac{-100 + 350}{1600} = \frac{10}{64} \text{ A}$$

$$i_3 = \frac{\begin{vmatrix} 20 & -10 & 5 \\ -10 & 20 & 0 \\ 0 & -5 & -1 \end{vmatrix}}{1600} = \frac{-400 + 250 + 100}{1600} = -\frac{2}{64} \text{ A}$$

$$v_1 = 10(i_1 - i_2) = 10\left(\frac{21}{64} - \frac{10}{64}\right) = \frac{110}{64} = 1.72 \text{ V}$$

$$v_2 = 5(i_2 - i_3) = 5\left(\frac{10}{64} + \frac{2}{64}\right) = \frac{60}{64} = 0.94 \text{ V}$$

2-28

Determine the value of the current i_x using the mesh analysis method

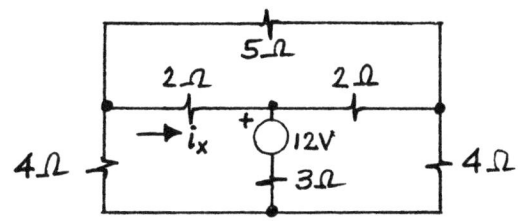

****

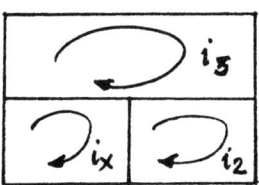

- ASSUMED CURRENTS ARE SHOWN.
- KVL EQS AROUND THE THREE PATHS ARE:

$$9i_x - 3i_2 - 2i_3 = -12$$
$$-3i_x + 9i_2 - 2i_3 = 12$$
$$-2i_x - 2i_2 + 9i_3 = 0$$

$$i_x = \frac{\begin{vmatrix} -12 & -3 & -2 \\ 12 & 9 & -2 \\ 0 & -2 & 9 \end{vmatrix}}{\begin{vmatrix} 9 & -3 & -2 \\ -3 & 9 & -2 \\ -2 & -2 & 9 \end{vmatrix}} = \frac{-552}{552} = -1.0 \text{ A}$$

2-29

Write the mesh equations for the circuit and find the current through the 5 ohm resistor in determinant form. Do not solve for the current.

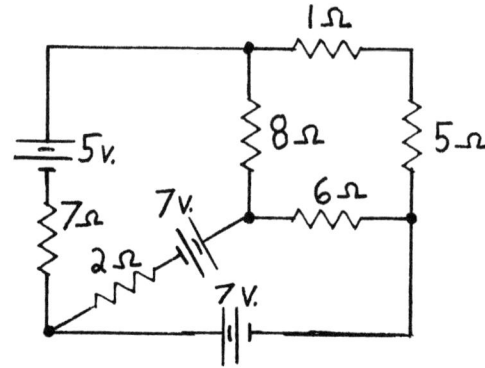

****

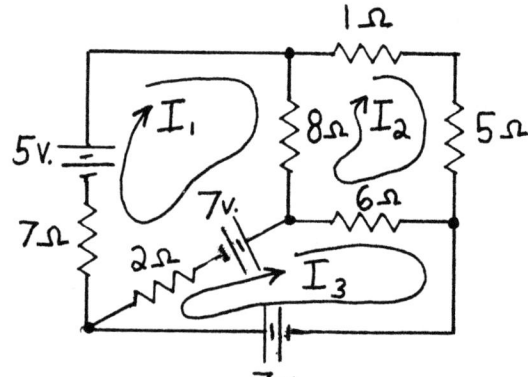

$$\sum_{mesh} V_{rises} \begin{pmatrix} \text{Voltage} \\ \text{Sources} \end{pmatrix} = \sum_{mesh} V_{drops} \begin{pmatrix} \text{other} \\ \text{elements} \end{pmatrix}$$

1ST MESH $\quad 5-7 = I_1(8+2+7) + I_2(-8) + I_3(-2)$

2ND MESH $\quad 0 = I_1(-8) + I_2(8+1+5+6) + I_3(-6)$

3RD MESH $\quad 7+7 = I_1(-2) + I_2(-6) + I_3(2+6)$

SIMPLIFIED:

$$-2 = I_1(17) + I_2(-8) + I_3(-2)$$

$$0 = I_1(-8) + I_2(20) + I_3(-6)$$

$$14 = I_1(-2) + I_2(-6) + I_3(8)$$

$$I_{5\Omega} = I_2 = \frac{\begin{vmatrix} 17 & -2 & -2 \\ -8 & 0 & -6 \\ -2 & 14 & 8 \end{vmatrix}}{\begin{vmatrix} 17 & -8 & -2 \\ -8 & 20 & -6 \\ -2 & -6 & 8 \end{vmatrix}} \text{ AMPS}$$

2-30

First determine the mesh currents, and then determine the node c voltage.

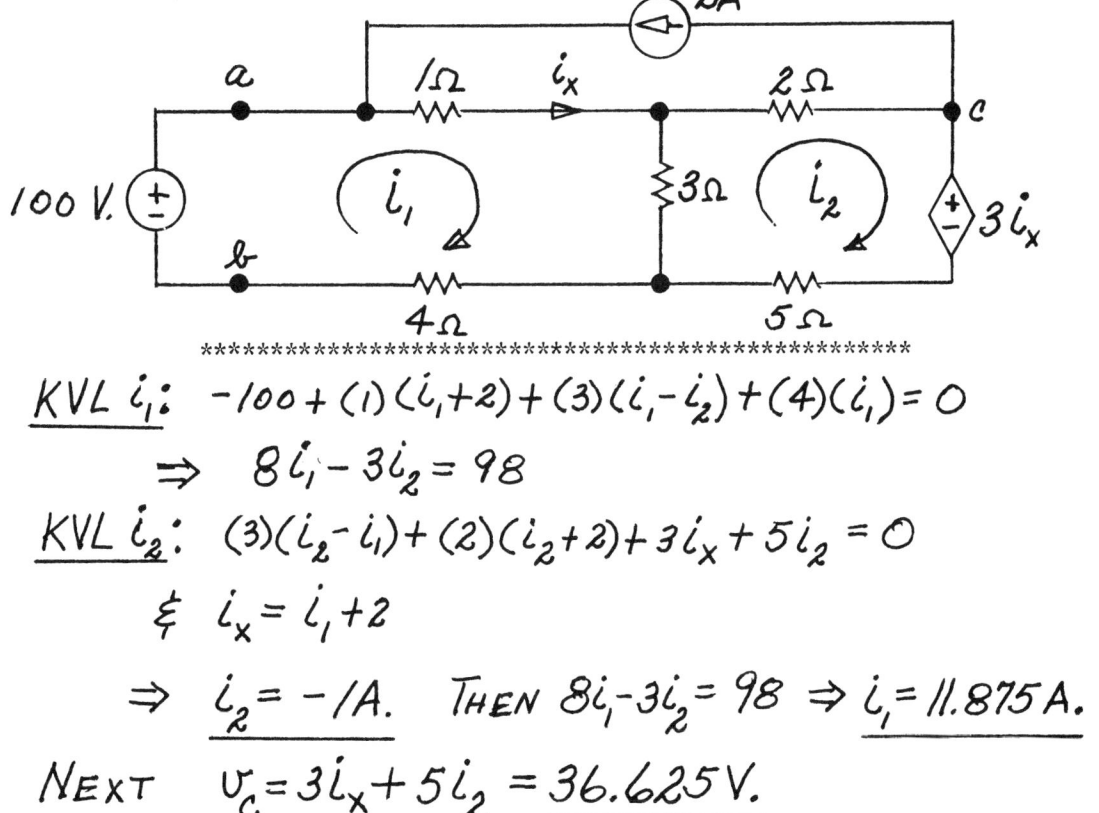

KVL i_1: $-100 + (1)(i_1 + 2) + (3)(i_1 - i_2) + (4)(i_1) = 0$
$\Rightarrow 8i_1 - 3i_2 = 98$

KVL i_2: $(3)(i_2 - i_1) + (2)(i_2 + 2) + 3i_x + 5i_2 = 0$
  & $i_x = i_1 + 2$

$\Rightarrow \underline{i_2 = -1 A.}$ THEN $8i_1 - 3i_2 = 98 \Rightarrow \underline{i_1 = 11.875 A.}$

NEXT $v_c = 3i_x + 5i_2 = \underline{36.625 V.}$

2-31

Write the mesh equations required for determining the mesh currents in the following circuit.

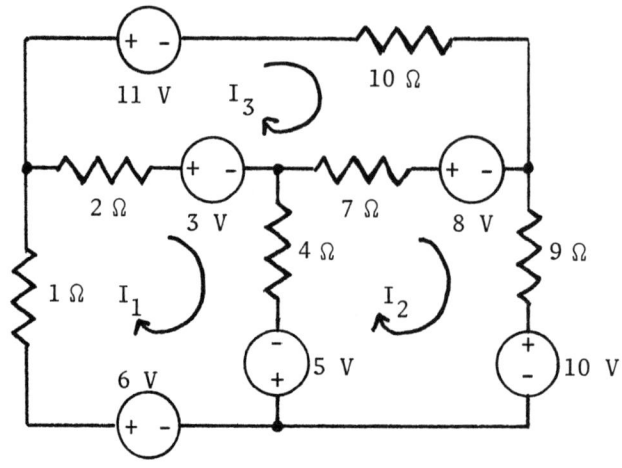

***

The sum of the voltage rises in a clockwise direction from voltage sources are $6-3+5 = 8$ V for mesh 1, $-5-8-10 = -23$ V for mesh 2, and $8+3-11 = 0$ V for mesh 3. The mesh self-resistances are

$$1+2+4 = 7 \ \Omega \text{ for mesh 1}$$
$$4+7+9 = 20 \ \Omega \text{ for mesh 2}$$
and
$$2+10+7 = 19 \ \Omega \text{ for mesh 3}$$

The mutual resistances are 4 Ω for meshes 1 and 2, 2 Ω for meshes 1 and 3, and 7 Ω for meshes 2 and 3. So, the mesh equations are

$$7I_1 - 4I_2 - 2I_3 = 8$$
$$-4I_1 + 20I_2 - 7I_3 = -23$$
$$-2I_1 - 7I_2 + 19I_3 = 0$$

or

$$\begin{bmatrix} 7 & -4 & -2 \\ -4 & 20 & -7 \\ -2 & -7 & 19 \end{bmatrix} \begin{bmatrix} I_1 \\ I_2 \\ I_3 \end{bmatrix} = \begin{bmatrix} 8 \\ -23 \\ 0 \end{bmatrix}$$

3
NETWORK THEOREMS

LINEARITY AND SUPERPOSITION

------3-1

Find the power delivered to the 3 ohm resistor using superposition.

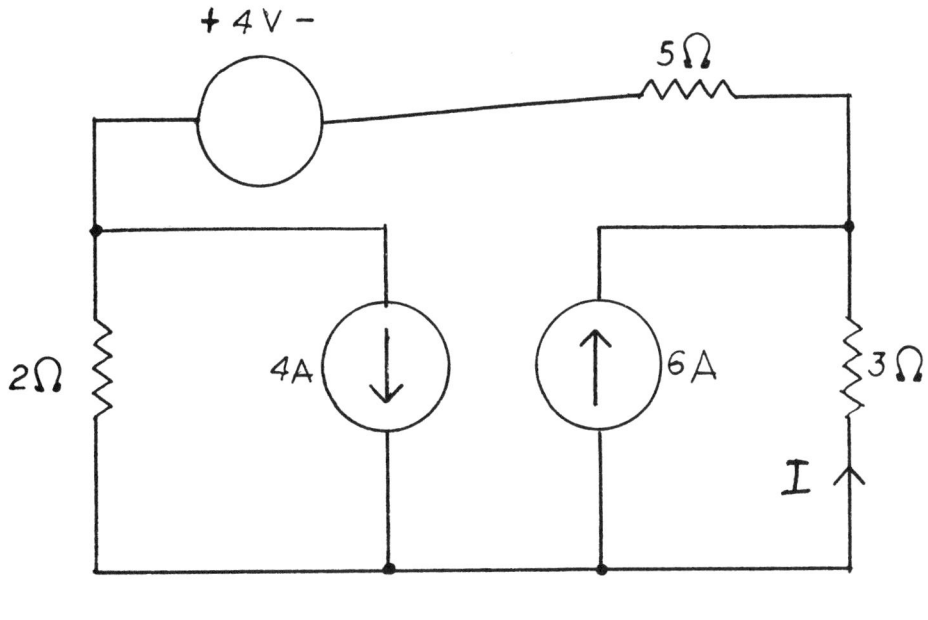

Find current due to each source separately. With the current sources dead they are open circuits. The three resistors are now in series and using ohm's law.

$$I_1 (4 \text{ Volts}) = \frac{4}{2+3+5} = 0.4 \text{ Amps.}$$

With the six ampere source dead it is an open circuit and the dead voltage source is a short circuit. The five ohm and three ohm resistors are now in series and they are in parallel with the 2 ohm resistor. By current division.

$$I_2 (4 \text{ Amps}) = 4 \frac{2}{2+3+5} = 0.8 \text{ Amps.}$$

With the 4 ampere source dead the 5 ohm and 2 ohm resistors are now in series and they are in parallel with the three ohm resistor.

$$I_3 (6 \text{ Amps.}) = -6 \frac{7}{7+3} = -4.2 \text{ Amps.}$$

Note that this current is in the opposite direction from the other two.

$$I = I_1 + I_2 + I_3 = 0.4 + 0.8 - 4.2 = -3 \text{ Amps.}$$
$$P = I^2 R = (-3)^2 \times 3 = \underline{27 \text{ Watts.}}$$

3-2

Find the current I_L that flows through the resistor R_L using the Superposition Theorem.

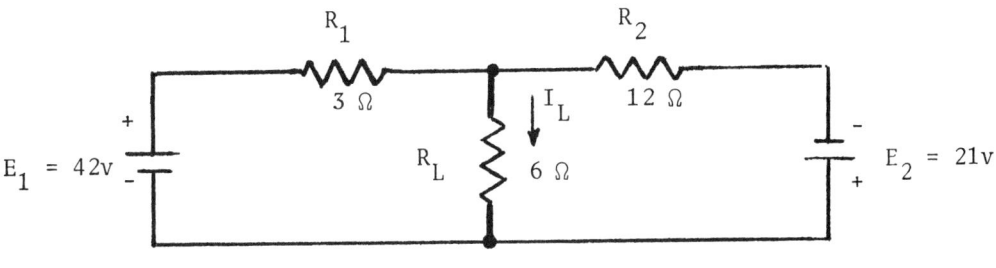

***

A) Consider the effects of E_1 (Short E_2)

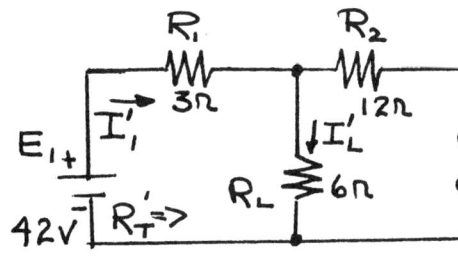

calculate total resistance
$$R_T' = R_1 + (R_2 \| R_L)$$
$$R_T' = 3\Omega + \frac{(12\Omega)(6\Omega)}{12\Omega + 6\Omega} = 7\Omega$$

Calculate total current
$$I_1' = \frac{E_1}{R_T'} = \frac{42V}{7\Omega} = 6a$$

Calculate I_L' (current divider rule)
$$I_L' = \frac{R_2}{R_2 + R_L} \times I_1' = \frac{12\Omega}{12\Omega + 6\Omega} \times 6a = 4a$$

B) Consider the effects of E_2 (Short E_1)

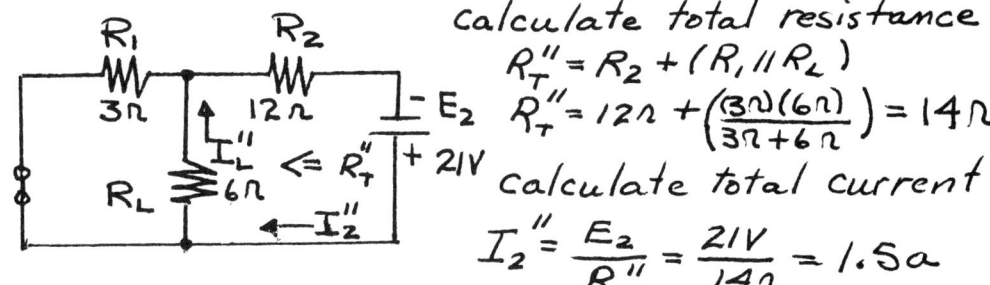

calculate total resistance
$$R_T'' = R_2 + (R_1 \| R_L)$$
$$R_T'' = 12\Omega + \left(\frac{(3\Omega)(6\Omega)}{3\Omega + 6\Omega}\right) = 14\Omega$$

calculate total current
$$I_2'' = \frac{E_2}{R_T''} = \frac{21V}{14\Omega} = 1.5a$$

Calculate I_L'' (current divider rule)
$$I_L'' = \frac{R_1}{R_1 + R_L} \times I_2'' = \frac{3\Omega}{3\Omega + 6\Omega} \times 1.5a = .5a$$

C) I_L is the algebraic sum of I_L' and I_L''
$$I_L = I_L'(\downarrow) - I_L''(\uparrow) = 4a(\downarrow) - .5a(\uparrow) = 3.5a(\downarrow)$$

3-3

Determine the current i by the method of superposition.

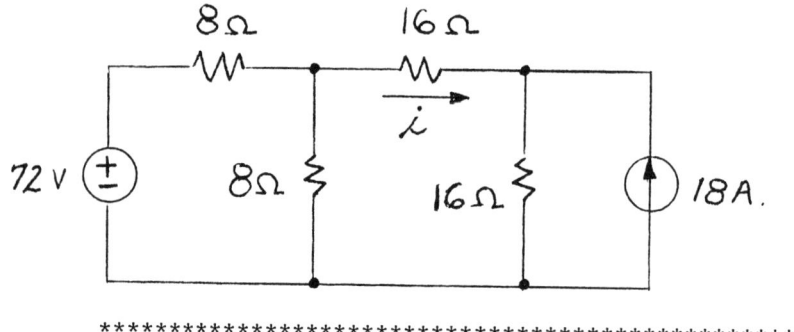

****

WITH THE 72 V. SOURCE ACTING ALONE, THE 18 A. SOURCE IS OPENED AND

$$i_1 = \frac{72}{8 + \frac{32(8)}{40}} = 5 A.$$

WITH THE 18 A. SOURCE ACTING ALONE AND THE 72 V SOURCE REPLACED BY A SHORT CIRCUIT, WE HAVE

$$i_2 = -\frac{16}{20 + 16}(18) = -8 A.$$

THEREFORE, BY SUPERPOSITION,

$$i = i_1 + i_2 = -3 A.$$

3-4

Find the time-varying component (sometimes called the ac component) and the average value (sometimes called the dc component) of the voltage drop across the 800 ohm resistor shown in the diagram.

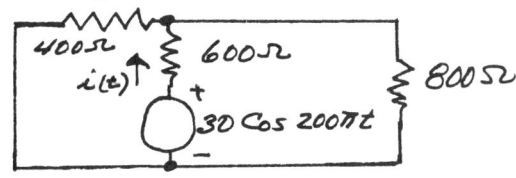

****

Applying superposition, the circuit for only the time-varying source is

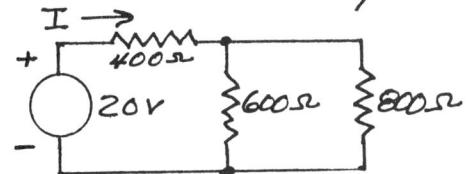

Combining the 400Ω and the 800Ω resistors:

$$R_{eq} = \frac{400 \times 800}{400 + 800} = 266.667 \, \Omega$$

KVL for the loop: $-30 \cos 200\pi t + 600 \, i(t) + 266.667 \, i(t) = 0$

$$i(t) = 0.034615 \cos 200\pi t \quad A$$

$$v_{800\Omega}(t) = (266.667)(0.034615 \cos 200\pi t)$$
$$= 9.231 \cos 200\pi t \quad V$$

The circuit for only the time-independent source is

Combining the 600Ω and 800Ω

$$R_{eq} = \frac{600 \times 800}{600 + 800} = 342.857 \, \Omega$$

KVL for the loop: $-20 + 400 I + 342.857 I = 0$

$$I = 0.02692 \quad A$$

$$V_{800\Omega} = (342.857)(0.02692) = 9.231 \, V$$

3-5

Using the Superposition Theorem, find the current flowing through the 6 ohm resistor.

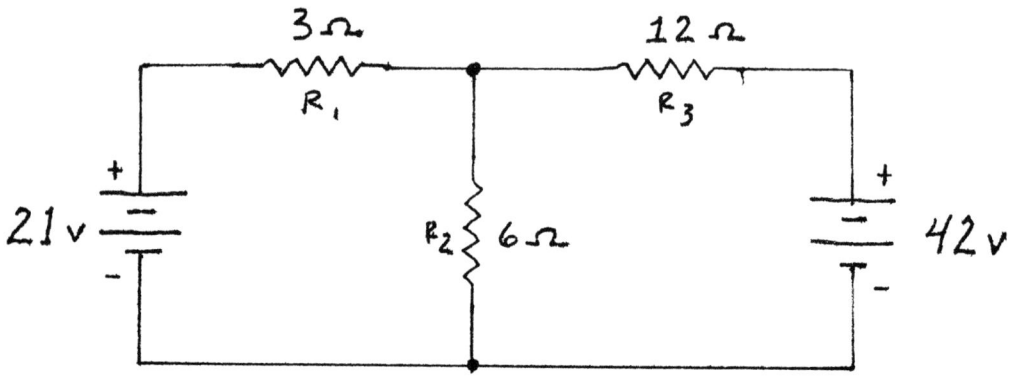

** * ** *** *** * ***** ** ** *** *** ** *** ********* ****** ***

STEP A:

(source reduced to 0 volts)

$R_T = 7\,\Omega$

$I = \dfrac{E}{R}$

$I = \dfrac{21v}{7\,\Omega}$

$I = 3\,A$

Current divider:

$I_{6\Omega} = \dfrac{I \cdot R_3}{R_2 + R_3}$

$I_{6\Omega} = \dfrac{3 \cdot 12}{6 + 12}$

$\underline{I_{6\Omega} = 2\,\text{Amps}}$

(first source)

STEP B:

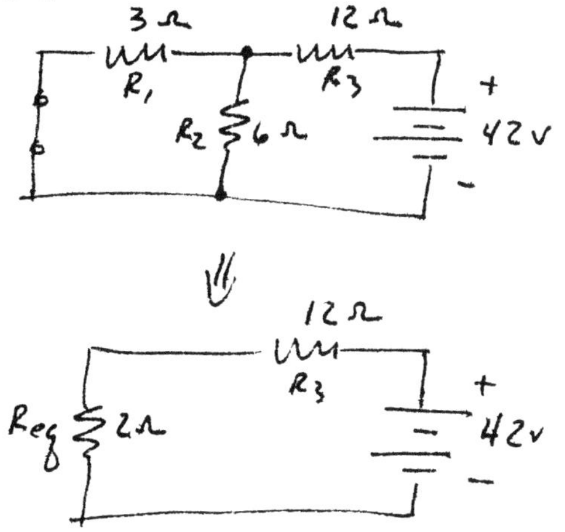

[Circuit diagram: $R_1 = 3\Omega$ in series with parallel combination of $R_2 = 6\Omega$ and ($R_3 = 12\Omega$ in series with 42V source)]

⇓

[Simplified circuit: $R_{eq} = 2\Omega$ in series with $R_3 = 12\Omega$ and 42V source]

$\underline{R_T = 14\Omega}$

$I = \dfrac{E}{R}$

$I = \dfrac{42V}{14\Omega}$

$\underline{I = 3 \text{ Amps}}$

Current divider:

$I_{6\Omega} = \dfrac{R_1 \cdot I}{R_1 + R_2}$

$I_{6\Omega} = \dfrac{3 \cdot 3}{3 + 6}$

$\underline{I_{6\Omega} = 1 \text{ Amp}}$

(second source

STEP C:

TAKE ALGEBRAIC SUM OF CURRENTS.
IN THIS CASE, CURRENTS FROM THE SOURCES
ARE AIDING, ∴

$\underline{\text{TOTAL } I_{6\Omega} = 3 \text{ Amps}}$

3-6

For the circuit shown, use superposition to find the current in resistance R_1.

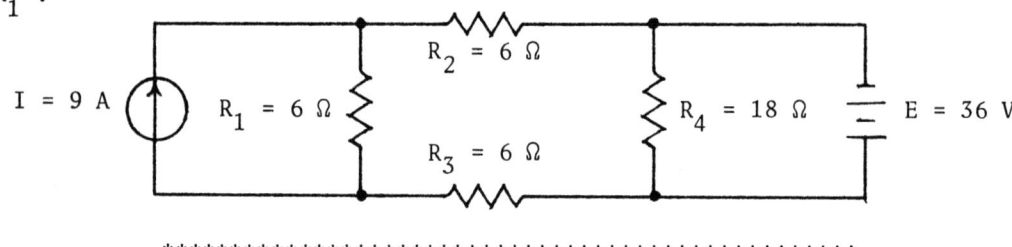

***

1) Circuit for current source:

$I = 9A$, $R_1 = 6\Omega$, I_{11}, $R_2 = 6\Omega$, $R_3 = 6\Omega$

Note that R_4 is shorted out!

$R_{23} = R_2 + R_3 = 12\Omega$.

$$R_{T1} = R_1 // R_{23} = \frac{6\Omega \cdot 12\Omega}{6\Omega + 12\Omega} = 4\Omega.$$

$$I_{11} = I \frac{R_{T1}}{R_1} = 9A \cdot \frac{4\Omega}{6\Omega} = \underline{6A \downarrow}.$$

2) Circuit for voltage source:

open → $R_1 = 6\Omega$, $I_{12}\uparrow$, $R_2 = 6\Omega$, $R_4 = 18\Omega$, $R_3 = 6\Omega$, $E = 36V$

$R_{1-3} = R_1 + R_2 + R_3 = 18\Omega$.

$$I_{12} = \frac{E}{R_{1-3}} = \frac{36}{18} = \underline{2A \uparrow}$$

3) $I_1 \downarrow = I_{11} - I_{12} = 6A - 2A = \underline{4A}$. (down)

3-7

Determine the voltage V_{AB} by the method of superposition.

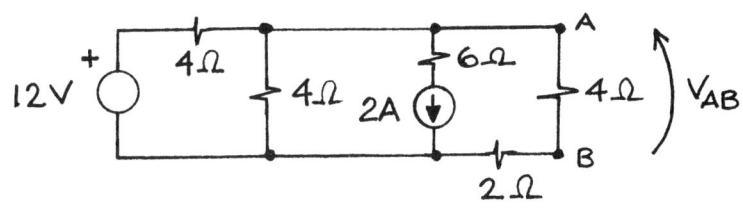

****

THE VOLTAGE SOURCE CIRCUIT. USE NODAL ANALYSIS TO FIND $(V_{AB})_V$

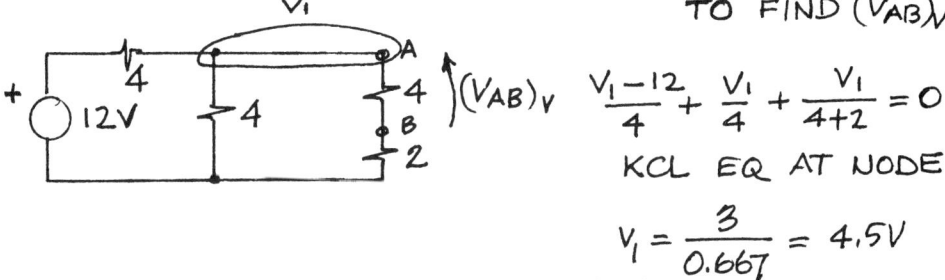

KCL EQ AT NODE:
$$\frac{V_1-12}{4}+\frac{V_1}{4}+\frac{V_1}{4+2}=0$$

$$V_1 = \frac{3}{0.667} = 4.5V$$

APPLYING VOLTAGE DIVIDER RULE

$$(V_{AB})_V = \left(\frac{4}{4+2}\right)V_1 = \frac{2}{3} \times 4.5 = 3V$$

THE CURRENT SOURCE CIRCUIT. USE NODAL ANALYSIS TO FIND $(V_{AB})_C$

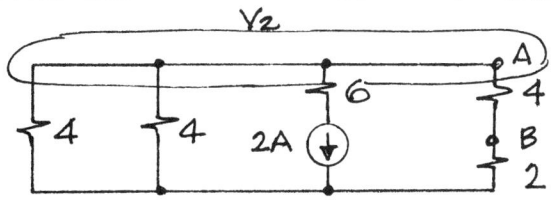

KCL EQ AT NODE: $\quad \frac{V_2}{4}+\frac{V_2}{4}+2+\frac{V_2}{6}=0$

$$V_2 = -\frac{2}{0.667} = -3V$$

APPLYING VOLTAGE DIVIDER RULE

$$(V_{AB})_C = \left(\frac{4}{4+2}\right)V_2 = \frac{2}{3}(-3) = -2V$$

$$V_{AB} = (V_{AB})_V + (V_{AB})_C = 3-2 = 1V$$

3-8

Solve for the current through the 5 ohm resistor by superposition.

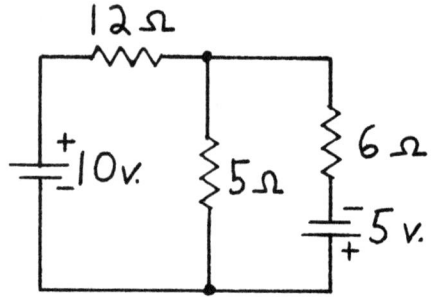

***

$R_T = 12 + \left(\frac{1}{5} + \frac{1}{6}\right)^{-1}$

$= 14.7\,\Omega$

$I_{T_1} = \frac{10}{14.7} = .679$ AMPS.

$I_1 = .679 \cdot \frac{6}{5+6}$

$= .370$ AMPS.

$R_T = 6 + \left(\frac{1}{5} + \frac{1}{12}\right)^{-1}$

$= 9.53\,\Omega$

$I_{T_2} = \frac{-5}{9.53} = -.525$ AMPS.

$I_2 = (-.525) \cdot \frac{12}{5+12}$

$= -.370$ AMPS.

$I_{5\Omega} = I_1 + I_2 = .370 - .370 = \underline{0 \text{ AMPS}}.$

3-9

Find the contribution to i_x from each source. Combine all the contributions to determine the total amount of i_x.

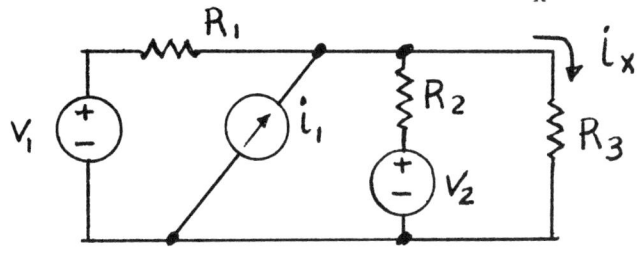

* *

SUPERPOSITION:

i_x DUE TO V_1

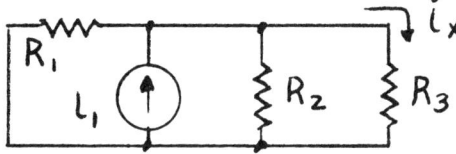

$$i_x = \frac{V_1}{R_1} \cdot \frac{R_1 \| R_2}{R_3 + (R_1 \| R_2)} = V_1 \frac{R_2}{R_3(R_1+R_2)+R_1 R_2} \text{ AMPS.}$$

i_x DUE TO i_1

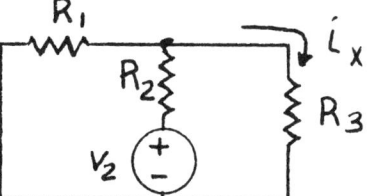

$$i_x = i_1 \frac{R_1 \| R_2}{R_3 + (R_1 \| R_2)} = i_1 \frac{R_1 R_2}{R_3(R_1+R_2)+R_1 R_2} \text{ AMPS.}$$

i_x DUE TO V_2

$$i_x = \frac{V_2}{R_2} \cdot \frac{R_1 \| R_2}{(R_1 \| R_2)+R_3} = V_2 \frac{R_1}{R_3(R_1+R_2)+R_1 R_2} \text{ AMPS.}$$

TOTAL i_x: $i_x = \dfrac{V_1 R_2 + i_1 R_1 R_2 + V_2 R_1}{R_3(R_1+R_2)+R_1 R_2}$ AMPS.

3-10

A. Use superposition to find the current in the 600 ohm resistor.
B. Find the Thevenin equivalent of the circuit connected to the 600 ohms.

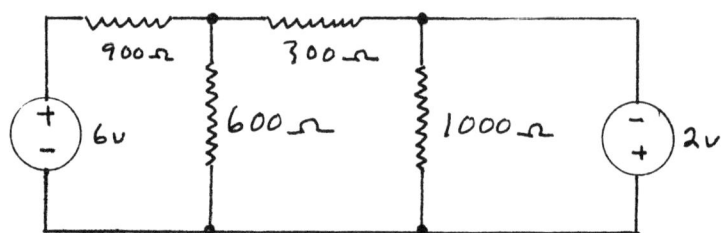

A.

We may ignore the $1000\,\Omega$ because a resistor across an _ideal_ voltage source has no effect on current in the rest of the circuit. To apply superposition, we short all voltage sources but one and figure contribution to the current from each source acting alone. First shorting the $2v$, we then have $300 \parallel 600$ in series with 900.

$R_{\parallel} = \dfrac{300 \times 600}{900} = 200\,\Omega$, so the $6v$ drives $900 + 200 = 1100\,\Omega$

Current through the $6v$ source is $\dfrac{6v}{1100\,\Omega} = \dfrac{60}{11}\,mA$

Apply current division, to get $I_{600} = \dfrac{300}{300+600} \times \dfrac{60}{11}\,mA = \dfrac{20}{11}\,mA$ (downward)

Shorting the $6v$ and ignoring $1000\,\Omega$, we have $900 \parallel 600$ in series with $300\,\Omega$

$R_{\parallel} = \dfrac{600 \times 900}{1500} = 360\,\Omega$, so current in $2v$ is $2v / (360 + 300)\,\Omega = (2/.66)\,mA$ flowing <u>down</u>

Again applying current division to get I_{600}

$I_{600} = \dfrac{2}{.66}\,mA \times \dfrac{900}{1500} = \dfrac{1800}{990} = \dfrac{20}{11}\,mA$ <u>upwards</u>

Thus, net value of $I_{600} = \boxed{0}$

B. Redrawing the key parts of the circuit:

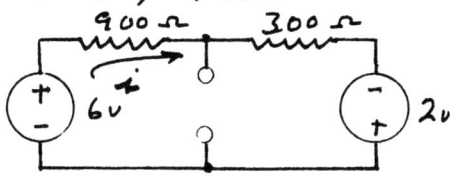

$$i = \frac{6v + 2v}{(900 + 300)\,\Omega} = \frac{8v}{1.2K} = 6\tfrac{2}{3}\,mA$$

$$V_{th} = 6v - 900\,\Omega \times 6\tfrac{2}{3}\,mA = 6v - 6v = \boxed{0}$$

Shorting all sources, resistance between terminals is $R_{11} = \frac{900 \times 300}{1200} = \boxed{225\,\Omega}$

3-11

Find V_o in the shown circuit. Assume that the operational amplifier is ideal.

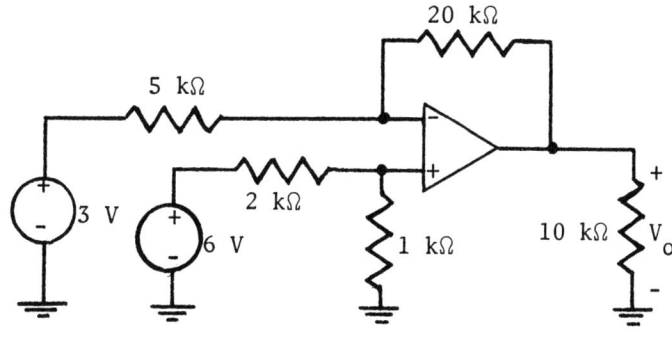

****

We can quickly obtain V_o by applying the superposition theorem to this circuit. With the 6-V source "killed," the op-amp circuit is an inverting amplifier with a gain of $-20/5 = -4$. And, with the 3-V source "killed," the circuit is a noninverting amplifier with a gain of $1 + 20/5 = 5$.

This second gain applies to v_+, the voltage at the non-inverting input terminal. By voltage division this voltage is

$$v_+ = \frac{1}{1+2} 6 = 2v$$

So, by superposition, $V_o = -4(3) + 5(2) = -2v.$

3-12

Determine the current in R using Superposition:

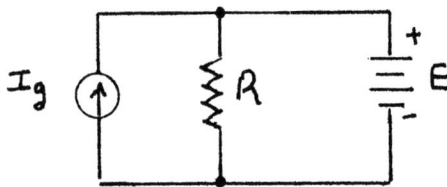

****

Replacing the battery with a short circuit, all current from the current source flows thru the short, and thus

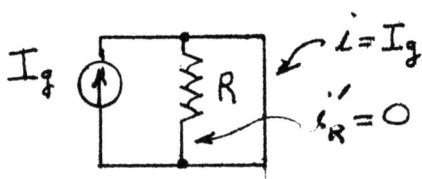

$i_R' = 0$

Replacing the current source by an open circuit,

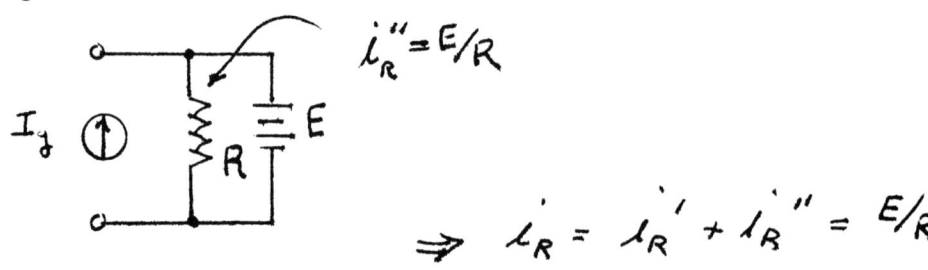

$$\Rightarrow i_R = i_R' + i_R'' = E/R$$

3-13

Use the principle of superposition to determine the voltage between nodes a & b.

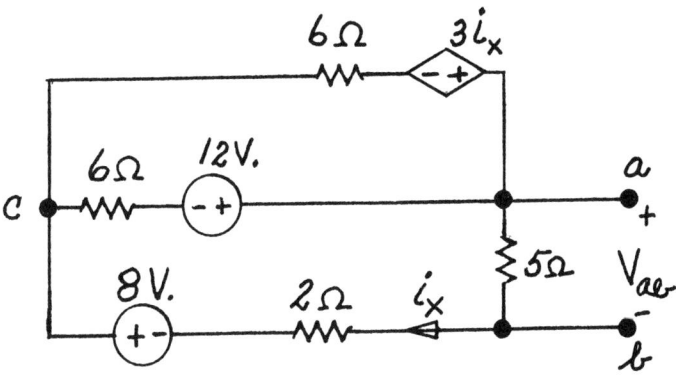

****

REPLACING THE 12V. SOURCE WITH ITS INTERNAL RESISTANCE AND PERFORMING SOURCE TRANSFORMATIONS WITH THE DEPENDENT SOURCE AND THE 6Ω RESISTORS YIELDS THE SINGLE LOOP CIRCUIT FROM WHICH V'_{ab} CAN BE OBTAINED.

$$5i'_x + 2i'_x - 8 + 3i'_x - \tfrac{3}{2}i'_x = 0$$

$$\Rightarrow i'_x = \tfrac{16}{17} A. \quad \& \quad V'_{ab} = 5i'_x$$

$$\Rightarrow V'_{ab} = \tfrac{80}{17} V.$$

REPLACING THE 8V. SOURCE AND PERFORMING SOURCE TRANSFORMATIONS AGAIN YIELDS A SINGLE LOOP CIRCUIT FROM WHICH V''_{ab} CAN BE OBTAINED.

$$5i''_x + 2i''_x + 3i''_x - 6 - \tfrac{3}{2}i''_x = 0$$

$$\Rightarrow i''_x = \tfrac{12}{17} A. \quad \& \quad V''_{ab} = 5i''_x = \tfrac{60}{17} V.$$

$$V_{ab} = V'_{ab} + V''_{ab} = \underline{\tfrac{140}{17} V.}$$

3-14

When V=5v and I=2A, i_x=1; when V=5v and I=-2A, i_x=0.
Find i_x when V=6v and I=3A.

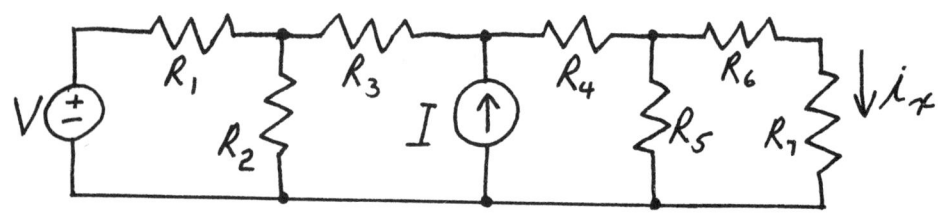

****

This is a linear circuit, therefore superposition holds. i_x is proportional to V and to I, with constants of proportionality C_v and C_i. This can be expressed by the two equations below.

$$5 C_v + 2 C_i = 1$$
$$5 C_v - 2 C_i = 0$$

Solving these two equations we obtain
$$C_v = 0.1 \quad C_i = 0.25$$

Using these values we find

$$i_x = 6(0.1) + 3(0.25) = \underline{1.35 A}$$

THE SUBSTITUTION THEOREM

---3-15

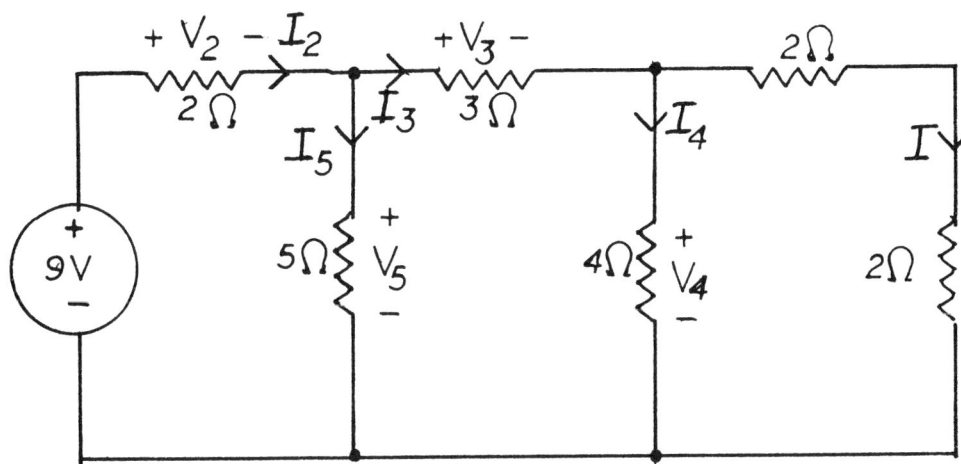

Assume a value of I, work back to the circuit input and using the linearity principle, calculate the correct value of I. Show work by calculating values of V_4, I_4, I_3, V_3, V_5, I_2, I_5, and V_2 for the assumed value of I.

****

Assume $I = 1$ Ampere. The two 2 ohm resistors to the right are in series.
$R = 2 + 2 = 4$ ohms
Then: $V_4 = 4I = 4 \times 1 = 4$ Volts — Ohm's Law
$I_4 = V_4/4 = 4/4 = 1$ Amp. — Ohm's Law
$I_3 = I + I_4 = 1 + 1 = 2$ Amps. — Current Law
$V_3 = 3 I_3 = 3 \times 2 = 6$ Volts — Ohm's Law
$V_5 = V_3 + V_4 = 6 + 4 = 10$ Volts — Voltage Law
$I_5 = V_5/5 = 10/5 = 2$ Amps. — Ohms Law
$I_2 = I_3 + I_5 = 2 + 2 = 4$ Amps — Current Law
$V_2 = 2 I_2 = 2 \times 4 = 8$ Volts — Ohm's Law
$V_{(INPUT)} = V_2 + V_5 = 8 + 10 = 18$ Volts
$I = 9/18 \times 1 = 0.5$ Amps — Proportionality

THEVENIN'S AND NORTON'S THEOREMS

3-16

When the switch is closed, I = 8A. Predict the value of v when the switch is open.

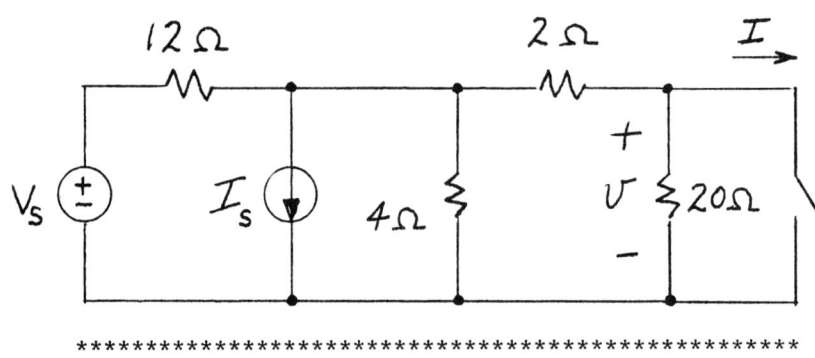

***

$I = 8A = I_{sc}$

THE SHORT CIRCUIT CURRENT IS 8 A.

THE THEVENIN RESISTANCE LOOKING AWAY FROM THE SWITCH IS CLEARLY

$$R_{th} = \{[(12 \| 4) + 2] \| 20\} = 4\,\Omega$$

THEREFORE, WHEN THE SWITCH IS OPEN, THE OPEN CIRCUIT VOLTAGE IS

$$v = I_{sc} R_{Th}$$
$$= (8)(4)$$

OR $v = 32\,V.$

3-17

Find the Thevenin equivalent circuit at terminals A, B.

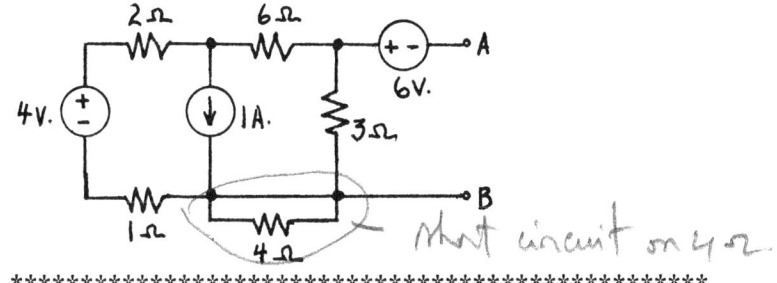

short circuit on 4Ω

****

DEAD NETWORK (FOR R_{th}):

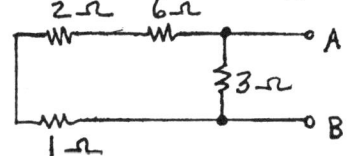

$R_{th} = \dfrac{3(9)}{3+9} = \dfrac{27}{12} = \dfrac{9}{4} \, \Omega$

REDUCE NETWORK (FOR $V_{OPEN\ CIRC.}$):

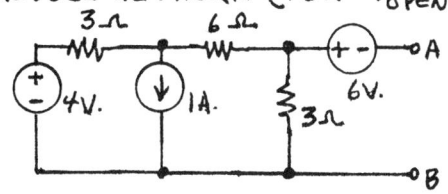 ⇒ 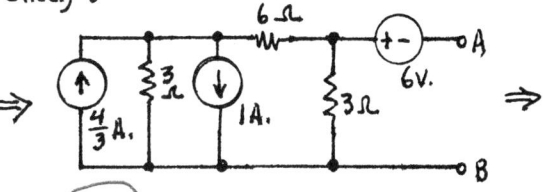 ⇒

⇒ 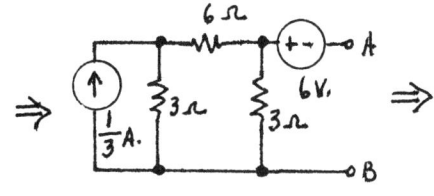 ⇒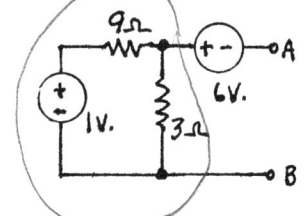

HERE, $V_{AB} = V_{OC}$
$= -6 + \dfrac{3}{3+9}(1)$ voltage divider
$= -\dfrac{23}{4}\ V.$

FINAL CIRCUIT:

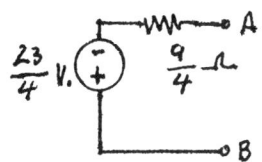

3-18

Find the Thevenin or output resistance R_{out} "looking into" terminals a,b of the shown circuit, which includes a model for the 741 op amp.

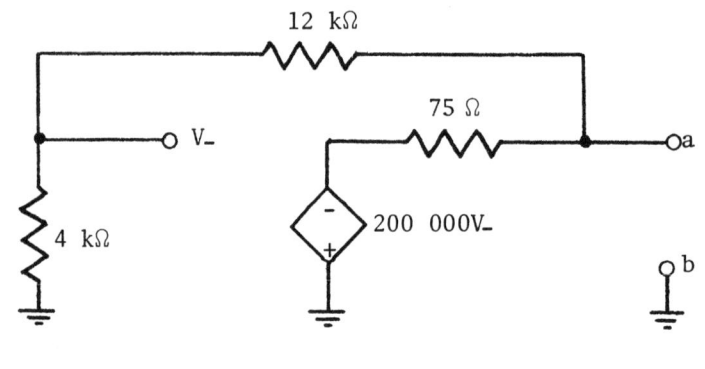

***

To find the output resistance, we must apply a source across terminals a, b, and find the resistance "seen" by this source, which is the ratio of the source voltage to the source current. We will arbitrarily apply a voltage source V_s as follows:

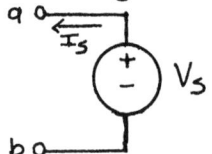

Then by voltage division,
$$V_- = \frac{4}{4+12} V_s = 0.25 V_s$$
and so
$$200\,000\, V_- = 50\,000\, V_s$$

The source current I_s is equal to the sum of the currents to the left through the 12-kΩ and 75-Ω resistors:
$$I_s = \frac{V_s}{12\,000 + 4000} + \frac{V_s - (-50\,000 V_s)}{75}$$

Clearly, the first term on the right-hand side is negligible compared to the second term. Also negligible is the V_s in the numerator of the second term. So,
$$R_{out} = \frac{V_s}{I_s} = \frac{75}{50\,000} = 0.0015 = 1.5\,m\Omega$$

Note that because of the negative feedback of the circuit, the output resistance of $1.5\ m\Omega$ of the op-amp circuit is much less than the $75\text{-}\Omega$ output resistance of the op amp.

3-19

You have purchased five quarter-watt resistors and have designed/constructed the network as shown. Calculate the Thevenin equivalent resistance at every node-pair.

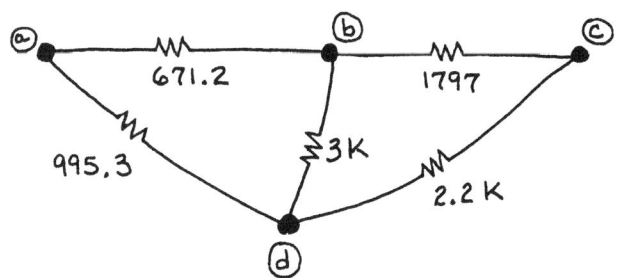

***

Constructing and solving the reference node d model:

$$\begin{bmatrix} \left(\frac{1}{671.2}+\frac{1}{995.3}\right) & -\frac{1}{671.2} & 0 \\ -\frac{1}{671.2} & \left(\frac{1}{671.2}+\frac{1}{3000}+\frac{1}{1797}\right) & -\frac{1}{1797} \\ 0 & -\frac{1}{1797} & \left(\frac{1}{1797}+\frac{1}{2200}\right) \end{bmatrix} \begin{bmatrix} V_a \\ V_b \\ V_c \end{bmatrix} = \begin{bmatrix} 0 \\ 0 \\ 0 \end{bmatrix}$$

$$\begin{bmatrix} V_a \\ V_b \\ V_c \end{bmatrix} = \begin{bmatrix} 702.2 & 504.6 & 277.7 \\ 504.6 & 844.9 & 465 \\ 277.7 & 465.0 & 1245 \end{bmatrix} \begin{bmatrix} 0 \\ 0 \\ 0 \end{bmatrix}$$

Six answers are needed. Three of the answers come from the main diagonal: $R_{TH}(a,d) = 702.2$; $R_{TH}(b,d) = 844.9$; $R_{TH}(c,d) = 1245$ ohms. Using the main and off diagonal terms: $R_{TH}(b,c) = \{844.9 - 465 + 1245 - 465\}$; $R_{TH}(a,c) = \{702.2 - 277.7 + 1245 - 277.7\}$; $R_{TH}(a,b) = \{\text{you do it}\}$.

3-20

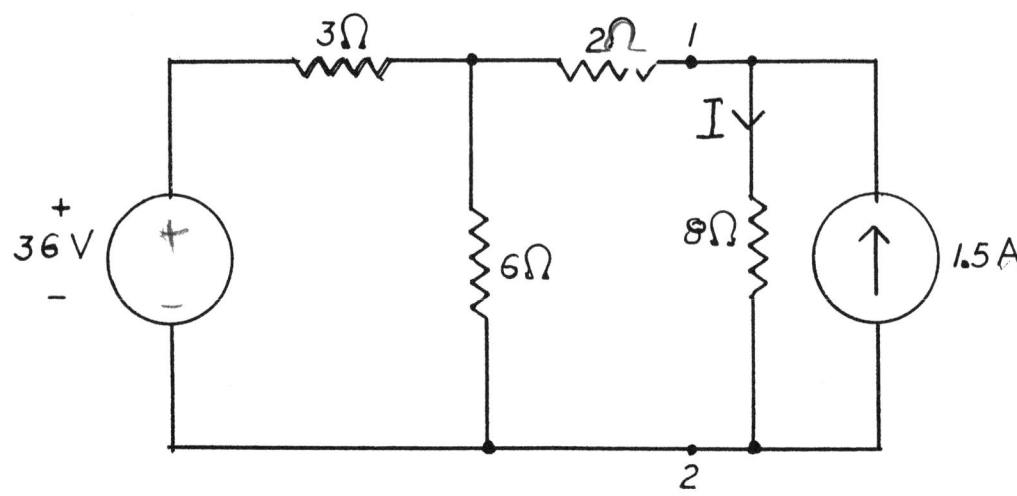

Find the Norton equivalent of the circuit to the left of terminals 1-2 and use it to find I.

***

Looking to the left at terminals 1-2 with the voltage source dead (short circuit), the 3 ohm and 6 ohm resistors are in parallel.

$\frac{3 \times 6}{3+6} = 2$ ohms. This two ohms is in series with the 2 ohm resistor.

$R(\text{Norton})$, $R_N = 2 + 2 = 4$ ohms

With terminals 1-2 open no current flows through the two ohm resistor and the open-circuit voltage, V_{oc}, is the voltage across the 6 ohm resistor. By voltage division

$$V_{oc} = 36 \times \frac{6}{6+3} = 24 \text{ Volts}$$

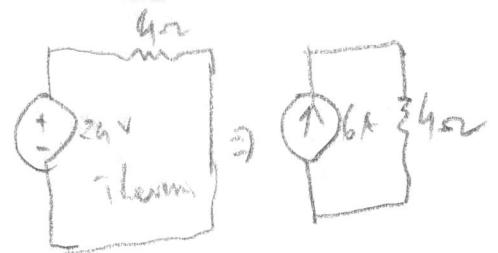

The short-circuit current is I_{SC}.

$$I_{SC} = \frac{V_{OC}}{R_N} = \frac{24}{4} = 6 \text{ Amps.}$$

Norton Equivalent is to left of 1-2.

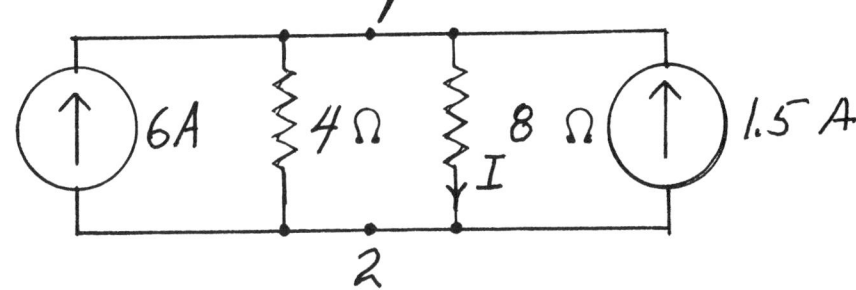

The two current sources are in parallel and add.

Source $I_S = 6 + 1.5 = 7.5$ Amps.

By current division.

$$I = I_S \frac{4}{4+8} = 7.5 \frac{4}{12} = \underline{2.5 \text{ Amps}}$$

3-21

For the circuit below determine the Thevenin equivalent circuit.

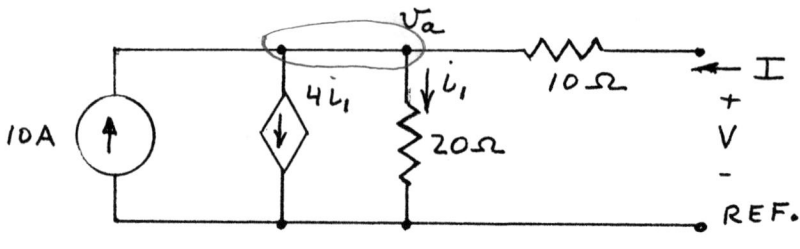

****

KCL AT V_a NODE:

$$10 - 4i_1 - i_1 + I = 0$$

$$i_1 = \frac{V_a}{20}$$

SUBST: $\quad 10 - 5\frac{V_a}{20} + I = 0 \quad$ ①

KVL: $\quad V = 10I + V_a \quad$ ②

ELIMINATE V_a FROM ① & ②
AND SOLVE FOR V:

$$V = 40 + 14I$$

∴ THEVENIN EQ CKT IS

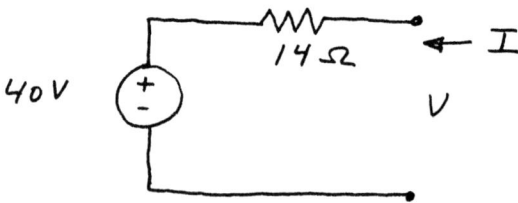

3-22

Find the current I_L that flows through the resistor R_L using Thevenin's Theorem with respect to terminals A and B.

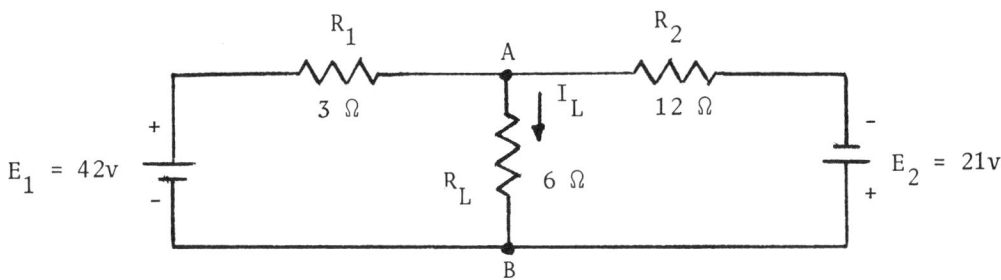

***

Calculate the Thevenin's Resistance R_{TH} (Short out both voltage supplies)

$$R_{TH} = R_1 \| R_2 = \frac{(3\Omega)(12\Omega)}{3\Omega + 12\Omega} = 2.4\Omega$$

Calculate the Thevenin's Voltage E_{TH} (open circuit voltage between terminals A and B) - (R_L removed)

$$I_T = \frac{E_1 + E_2}{R_1 + R_2} = \frac{42V + 21V}{3\Omega + 12\Omega} = 4.2a$$

$E_{TH} - V_2 + E_2 = 0$ (Kirchhoff's Voltage Rule)

$E_{TH} = -E_2 + V_2 = -21V + (4.2a)(12\Omega) = 29.4 V$

Thevenin's Equivalent Circuit

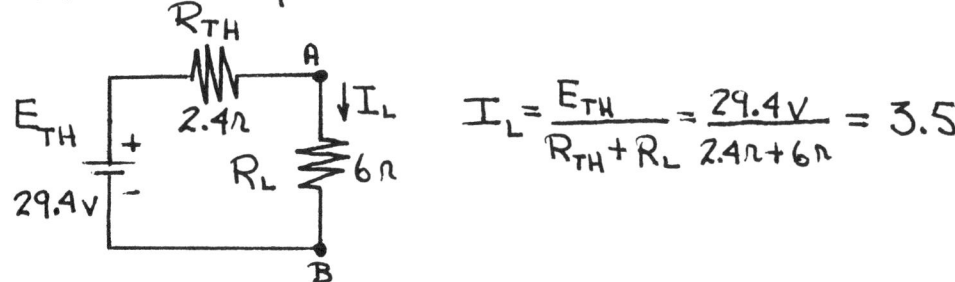

$$I_L = \frac{E_{TH}}{R_{TH} + R_L} = \frac{29.4V}{2.4\Omega + 6\Omega} = 3.5a$$

3-23

Find the Thevenin equivalent circuit for the circuit shown below at terminals a and b, all resistor values are in ohms.

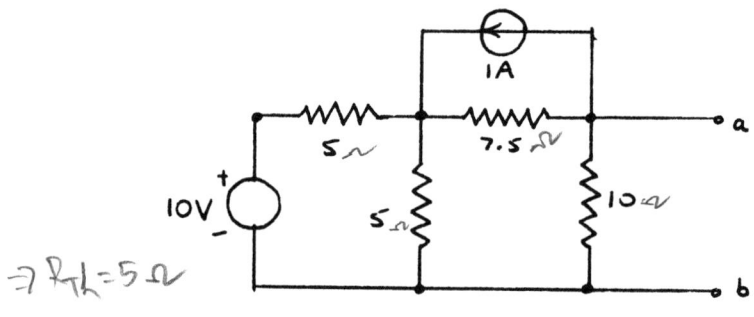

$\Rightarrow R_{Th} = 5\,\Omega$

****

Convert the current source in parallel with the 7.5 Ω resistor to a voltage source in series to get:

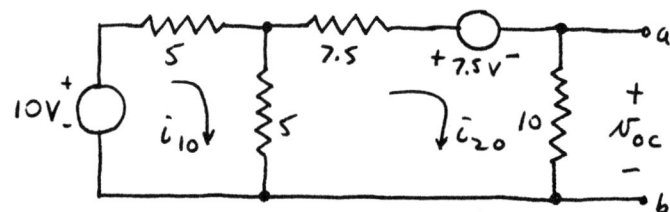

$-10 + 5i_{10} + 5(i_{10} - i_{20}) = 0$

$-10 + 5i_{10} + 5i_{10} - 5i_{20} = 0$

$-10 + 10i_{10} - 5i_{20} = 0 \quad (1)$

$10i_{10} - 5i_{20} = 10$

$-5i_{10} + 22.5 i_{20} = -7.5$

$5(i_{20} - i_{10}) + 7.5 i_{20} + 7.5 + 10 i_{20} = 0$

$5i_{20} - 5i_{10} + 7.5 i_{20} + 7.5 + 10 i_{20} = 0$

$22.5 i_{20} - 5 i_{10} - 7.5 = 0 \quad (2)$

$i_{20} = \dfrac{\begin{vmatrix} 10 & 10 \\ -5 & -7.5 \end{vmatrix}}{\begin{vmatrix} 10 & -5 \\ -5 & 22.5 \end{vmatrix}} = \dfrac{-75 + 50}{225 - 25} = \dfrac{-25}{200}$

$i_{20} = -\dfrac{1}{8}\,A \qquad v_{oc} = 10\, i_{20} = -\dfrac{5}{4} V = v_{th}$

Short the terminals a & b to get:

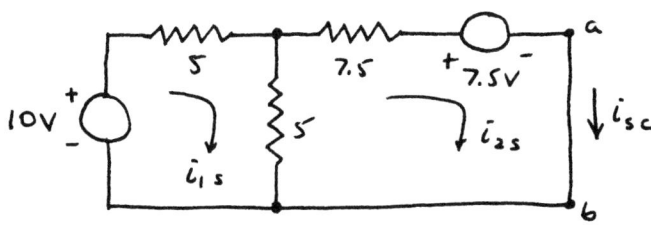

$$10\,i_{1s} - 5\,i_{2s} = 10$$

$$-5\,i_{1s} + 12.5\,i_{2s} = -7.5$$

$$i_{2s} = \frac{\begin{vmatrix} 10 & 10 \\ -5 & -7.5 \end{vmatrix}}{\begin{vmatrix} 10 & -5 \\ -5 & 12.5 \end{vmatrix}} = \frac{-75 + 50}{125 - 25} = -\frac{1}{4}\,A = i_{sc}$$

$$R_{th} = \frac{v_{oc}}{i_{sc}} = \frac{-5/4}{-1/4} = 5\,\Omega$$

∴ The Thevenin equivalent is:

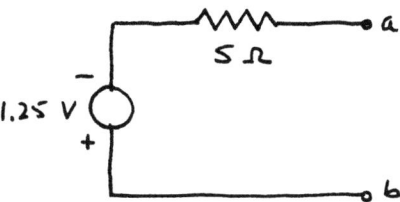

3-24

Find the Thevenin voltage and Thevenin resistance for each network shown.

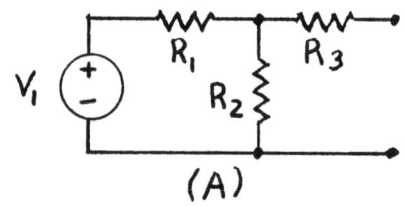
(A)

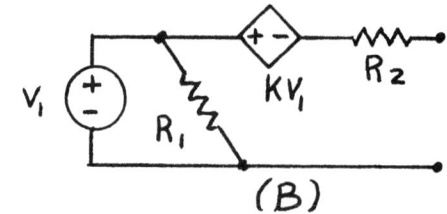
(B)

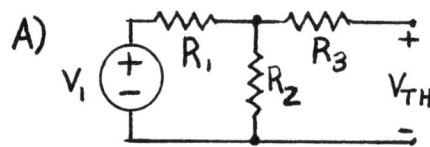
(C)

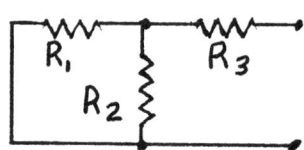
(D)

* *

A)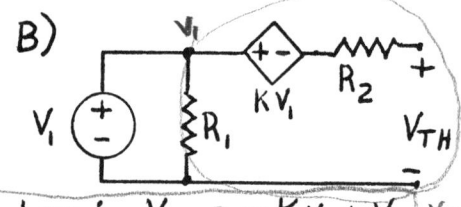

$V_{TH} = V_1 \dfrac{R_2}{R_1+R_2}$
(VOLTAGE DIVIDER)

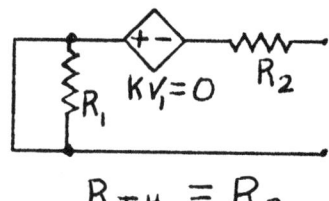

$R_{TH} = R_3 + (R_1 \| R_2)$

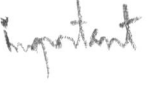
important

B) [figure: Loop: $V_{TH} = -KV_1 + V_1$ KVL]

[figure: $KV_1 = 0$, $R_{TH} = R_2$]

C) $V_{TH} = 0$
NO INDEPENDENT SOURCES IN THE NETWORK.

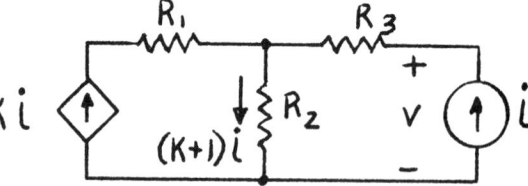

$V = iR_3 + i(K+1)R_2$
$R_{TH} = \dfrac{V}{i} = R_3 + (K+1)R_2$

D) $V_{TH} = 0$

NO INDEPENDENT SOURCES IN THE NETWORK.

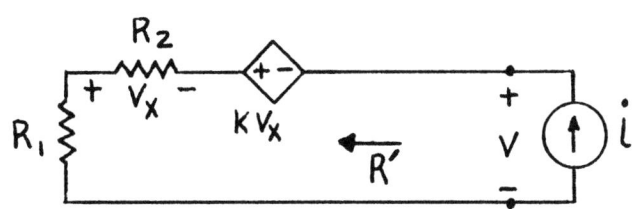

$$R_{TH} = R_3 \| R', \text{ WHERE } R' = \frac{V}{i}$$

Loop Eq.: $V = -KV_x + i(R_2 + R_1)$

Auxiliary Eq.: $-V_x = iR_2$

$$V = KiR_2 + i(R_2 + R_1)$$
$$R' = (K+1)R_2 + R_1$$
$$R_{TH} = R_3 \| [(K+1)R_2 + R_1]$$

3-25

Find the Thevenin equivalent circuit.

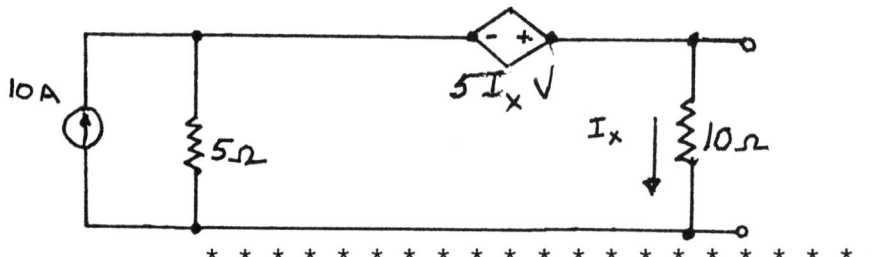

* * * * * * * * * * * * * * * * * * * *

Use source transformation,

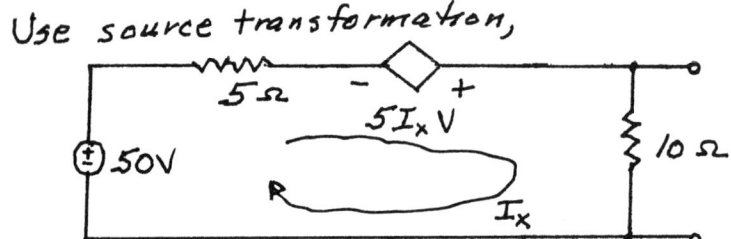

Writing loop eqn., $50 + 5I_x = 15 I_x$

$$I_x = 5 \text{ amps.}$$

$$V_{oc} = 10 I_x = 50 \text{ Volts}$$

For R_o, find I_{sc}

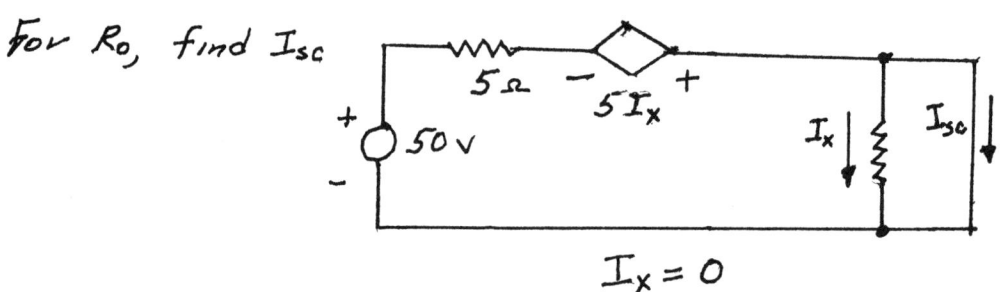

$$I_x = 0$$

Equation of circuit is $\quad 50 = 5 I_{sc}$

$$I_{sc} = 10 \text{ amps}$$

$$R_o = \frac{V_{oc}}{I_{sc}} = 5 \, \Omega$$

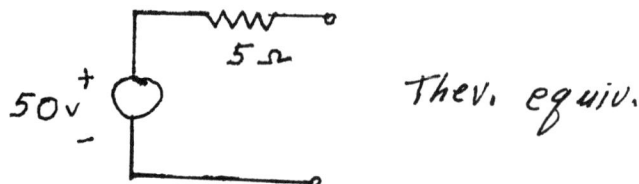

Thev. equiv.

3-26

Determine the Thevenin equivalent circuit to the right of terminals a & b.

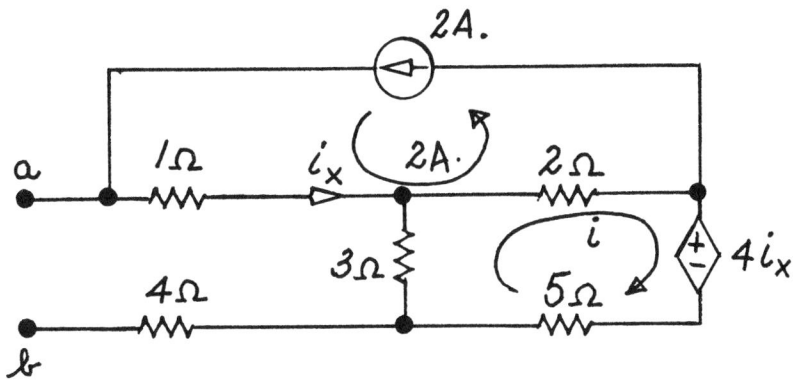

****

OPEN CIRCUIT VOLTAGE: $V_{oc} = V_t = V_{ab} = (1)(2) - (3)(i)$
$\qquad\qquad\qquad\qquad\qquad\qquad = 2 - (3)(-1.2)$

$i_x = 2A. \Rightarrow 4i_x = 8V.$

$\underline{V_{oc} = 5.6 V.}$

$\xi\ 5i + 3i + 2(i+2) + 8 = 0$
$\Rightarrow i = -1.2 A.$

THEVENIN EQUIVALENT IMPEDANCE: $R_t = \dfrac{V'_{ab}}{1}$

$V'_{ab} = (1)(1) + (3)(1 - i') + (4)(1)$
$\qquad = 8 - 3i'$
$\qquad = 8 - 3(-0.1)$
$V'_{ab} = 8.3 V.$

$\Rightarrow \underline{R_t = 8.3 \Omega}$

$5i' + (3)(i'-1) + 2i' + 4i'_x = 0$
$\xi\ i'_x = 1A. \Rightarrow i' = -0.1 A$

THEVENIN EQUIVALENT CIRCUIT:

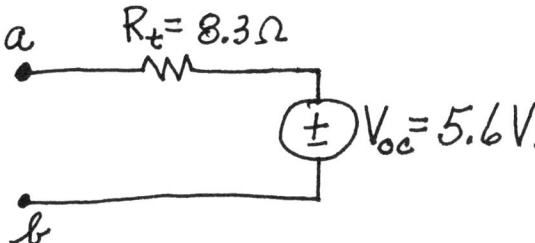

3-27

Reduce the circuit as seen at terminals a and b to the Norton equivalent circuit. All resistance values are in ohms.

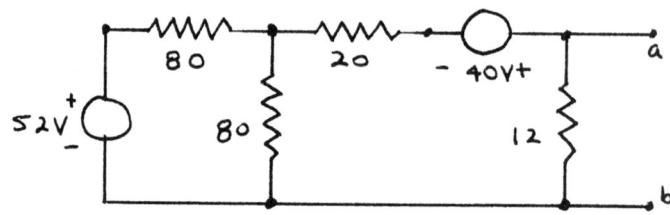

****

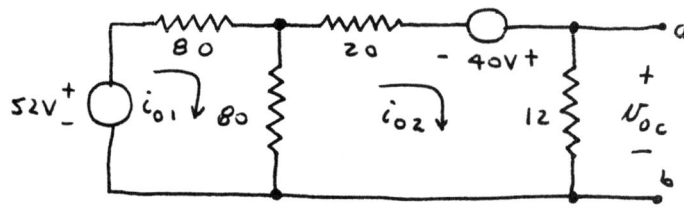

$$160 i_{o1} - 80 i_{o2} = 52$$

$$-80 i_{o1} + 112 i_{o2} = 40$$

$$i_{o2} = \frac{\begin{vmatrix} 160 & 52 \\ -80 & 40 \end{vmatrix}}{\begin{vmatrix} 160 & -80 \\ -80 & 112 \end{vmatrix}} = \frac{6400 + 4160}{17920 - 6400} = \frac{10560}{11520} \text{ A}$$

$$v_{oc} = 12 i_{o2} = \frac{10560}{960} = 11 \text{ V}$$

$$160 i_{s1} - 80 i_{s2} = 52$$

$$-80 i_{s1} + 100 i_{s2} = 40$$

$$i_{s2} = \frac{\begin{vmatrix} 160 & 52 \\ -80 & 40 \end{vmatrix}}{\begin{vmatrix} 160 & -80 \\ -80 & 100 \end{vmatrix}} = \frac{10560}{16000-6400} = \frac{10560}{960} = 1.1\,A$$

$$= i_{sc} = i_N$$

$$R_N = \frac{v_{oc}}{i_{sc}} = \frac{11}{1.1} = 10\,\Omega$$

∴ The Norton equivalent is:

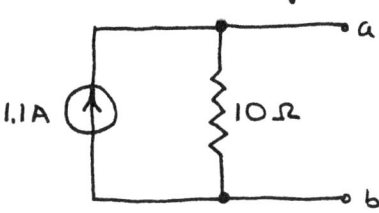

3-28

By applying Thevenin's theorem to the portion of the circuit shown between the points A and B, find the current through the load R_L.

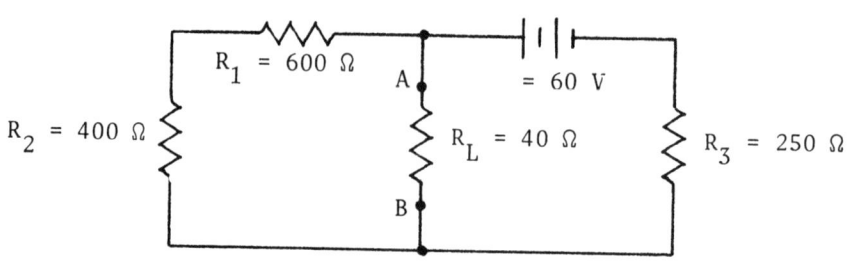

***

1) Find V_{Th}:

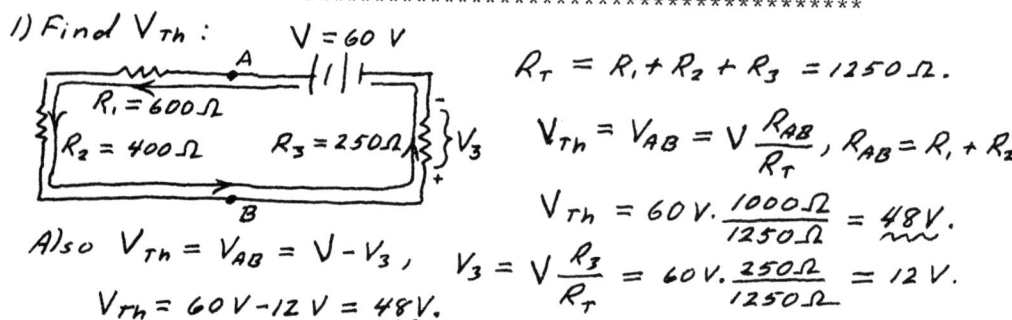

$R_T = R_1 + R_2 + R_3 = 1250\,\Omega$.

$V_{Th} = V_{AB} = V\dfrac{R_{AB}}{R_T}$, $R_{AB} = R_1 + R_2$

$V_{Th} = 60\,V \cdot \dfrac{1000\,\Omega}{1250\,\Omega} = 48\,V$.

Also $V_{Th} = V_{AB} = V - V_3$, $V_3 = V\dfrac{R_3}{R_T} = 60\,V \cdot \dfrac{250\,\Omega}{1250\,\Omega} = 12\,V$.

$V_{Th} = 60\,V - 12\,V = 48\,V$.

2) Find R_{Th}:

$R_{12} = R_1 + R_2 = 1000\,\Omega$.

$R_{Th} = R_{AB} = R_{12} \| R_3$

$R_{Th} = \dfrac{1000\,\Omega \cdot 250\,\Omega}{1000\,\Omega + 250\,\Omega} = 200\,\Omega$.

3) Thevenin Equivalent Circuit:

$V_{Th} = 48\,V$, $R_{Th} = 200\,\Omega$

4) Find I_L:

$V_{Th} = 48\,V$, $R_{Th} = 200\,\Omega$, $R_L = 40\,\Omega$

$R_T' = R_{Th} + R_L = 240\,\Omega$.

$I_L = \dfrac{V_{Th}}{R_T'} = \dfrac{48}{240} = .2\,A$.

3-29

Determine the values for the equivalent Norton circuit that would supply the same current to the resistance R_L as supplied by the 10V source.

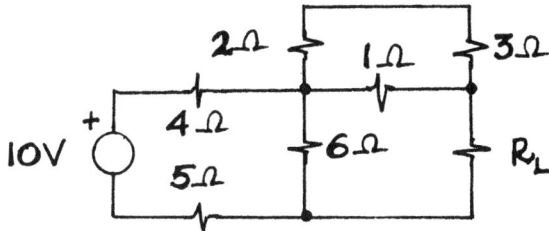

****

EQUIVALENT RESISTANCE CIRCUIT. COMBINE RESISTANCES

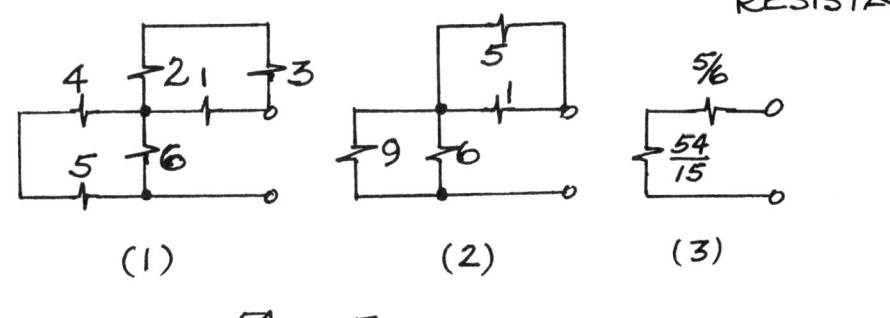

(1)   (2)   (3)

$$R_N = \frac{54}{15} + \frac{5}{6} = 4.43 \Omega$$

FIND THE SHORT CIRCUIT CURRENT, I_{sc},

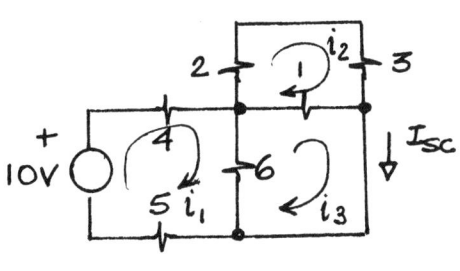

USING MESH ANALYSIS.

$i_3 = I_{sc}$

THE THREE KVL EQS ARE:

$15 i_1 \qquad -6 i_3 = 10$
$\qquad 6 i_2 - i_3 = 0$
$-6 i_1 - i_2 + 7 i_3 = 0$

$$i_3 = \frac{\begin{vmatrix} 15 & 0 & 10 \\ 0 & 6 & 0 \\ -6 & -1 & 0 \end{vmatrix}}{\begin{vmatrix} 15 & 0 & -6 \\ 0 & 6 & -1 \\ -6 & -1 & 7 \end{vmatrix}} = \frac{366}{399} = 0.9 A = I_{sc}$$

I_{sc} = NORTON CURRENT SOURCE = 0.9 A
R_N = NORTON RESISTANCE = 4.43 Ω

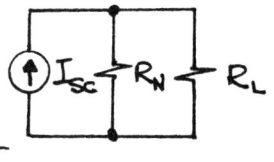

3-30

Find the Thevenin Equivalent Circuit for this circuit if R is considered as a load.

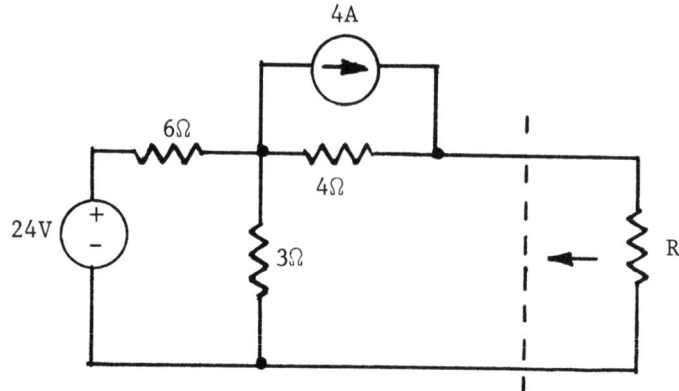

****

The Thevenin Resistance is Just The Resistance Seen Looking Back Into The Network With Voltage Source Shorted and The Current Source Opened.

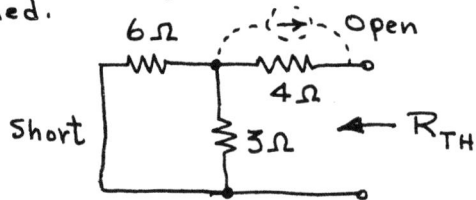

$$R_{TH} = 4 + 6//3 = 4 + 6 \cdot 3/(6+3) = 4 + 18/9 = 6\Omega$$

The Thevenin Voltage is the Open Circuit Voltage.

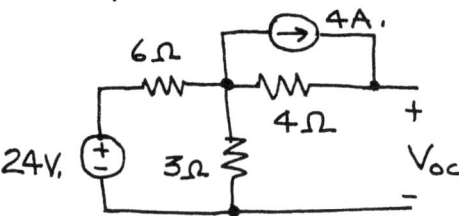

Since The 4A. Source And 4Ω Resistor Form An Isolated Mesh, Then

$$V_{OC} = V_{3\Omega} + V_{4\Omega} = \frac{3}{3+6}(24) + 4 \times 4$$

$$= 8 + 16 = 24 V,$$

$$V_{TH} = V_{OC} = 24 V.$$

3-31

Find the Thevenin equivalent for the circuit shown.

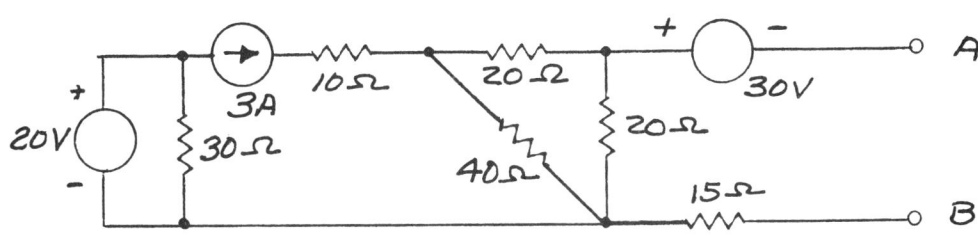

****

First, find the open-circuit voltage drop V_{AB}.
The 3A divides evenly between the 40Ω resistor and the two 20Ω resistors in series.
(Note that the current through the 30 volt source is zero.)

$I_{20\Omega} = 1.5A$

Hence $V_{20\Omega} = (20)(1.5) = 30V$

$V_{AB} = -30 + 30 = 0V$

Second, find the equivalent resistance R_{AB} when each source is replaced by its internal resistance.

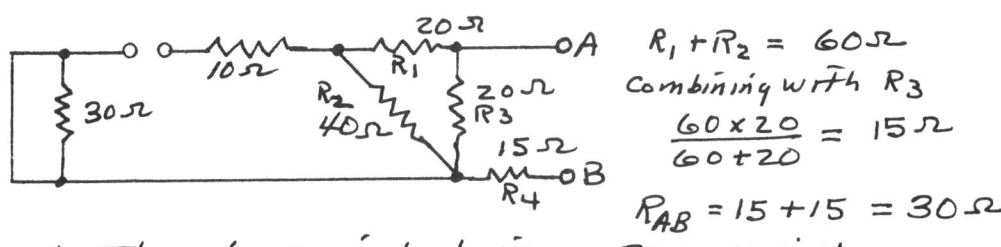

$R_1 + R_2 = 60\Omega$
Combining with R_3
$\dfrac{60 \times 20}{60 + 20} = 15\Omega$

$R_{AB} = 15 + 15 = 30\Omega$

The Thevenin equivalent is a 30Ω resistor

3-32

Find the Thevenin's equivalent circuit for the load resistor R.

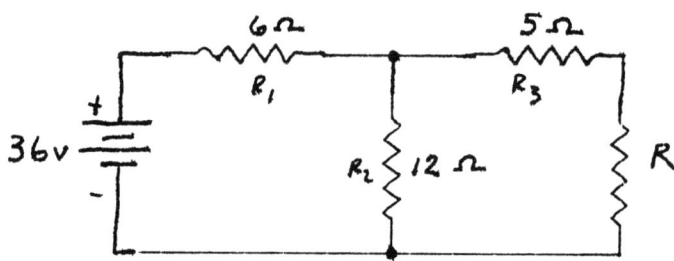

STEPS A and B: remove component for which equivalent is being found, mark terminals

STEP C: Find V_{TH} (voltage between marked terminals)

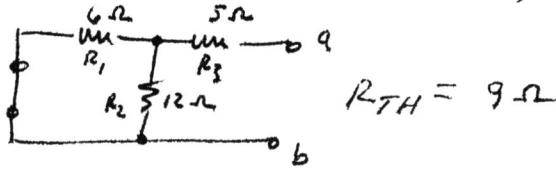

$$V_{TH} = \frac{R_2 \cdot E}{R_1 + R_2} = \frac{12 \cdot 36}{6 + 12} = 24 \text{ volts}$$

STEP D: Find R_{TH} (resistance you would measure between the marked terminals)

$R_{TH} = 9\,\Omega$

STEP E: Draw the equivalent circuit

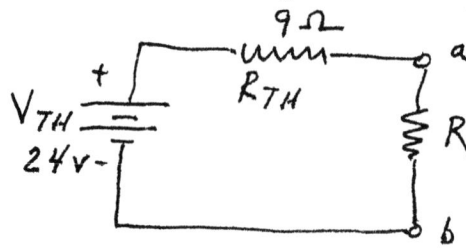

3-33

Find the Thevenin equivalent circuit at a-b.

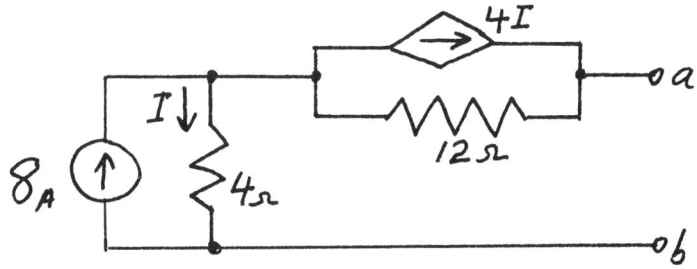

****

With a-b open circuit $I = 8A$. Current through $12\Omega = 4 \times 8 = 32A$
$V_{OC} = 8 \times 4 + 32 \times 12 = 416v$

Can Find R_{th} by two different methods

Method 1
$R_{th} = \dfrac{V_{OC}}{I_{SC}}$

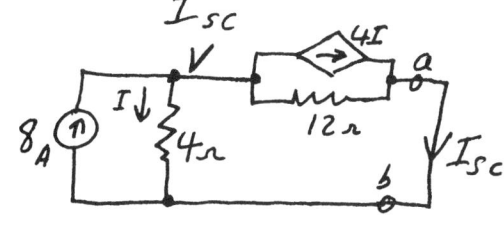

$\dfrac{V}{4} + \dfrac{V}{12} + 4\dfrac{V}{4} = 8$

$V = 6$

$I = \dfrac{6}{4} = \dfrac{3}{2}A$

$I_{SC} = 4(\dfrac{3}{2}) + \dfrac{6}{12} = 6\dfrac{1}{2}A$

$R_{th} = \dfrac{416}{6.5} = 64\Omega$

Method 2
set **Independent** source = 0
apply test current = 1_A

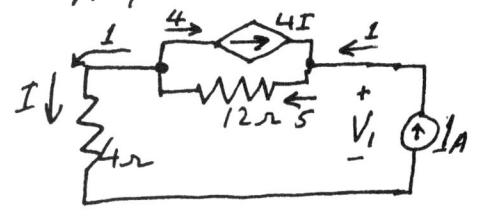

$I = 1$

$V_1 = 1 \times 4 + 5 \times 12 = 64v$

$R_{th} = \dfrac{64v}{1_A} = 64\Omega$

Checks

Thevenin Equivalent is 416_v

3-34

Find the Thevenin equivalent circuit at the terminals a-b.

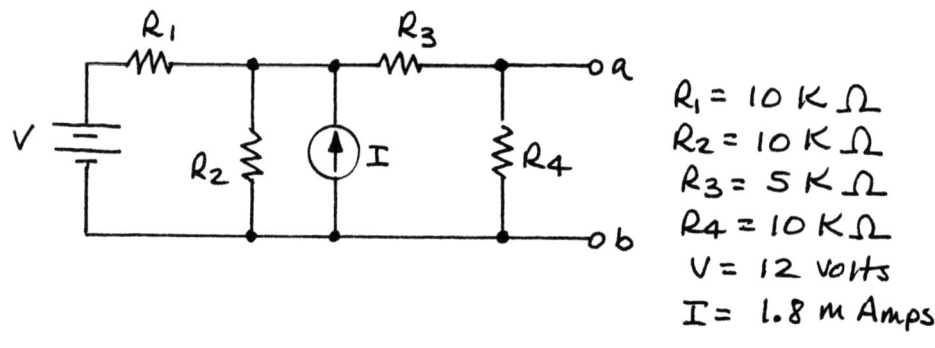

$R_1 = 10\ K\Omega$
$R_2 = 10\ K\Omega$
$R_3 = 5\ K\Omega$
$R_4 = 10\ K\Omega$
$V = 12\ volts$
$I = 1.8\ m\ Amps$

****

Solving for the open-circuit voltage V_{ab} by use of Nodal equations.

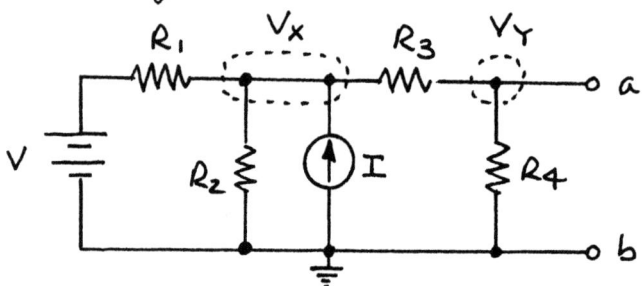

$V_Y = V_{ab} = V_{th} =$ Thevenin voltage

$$V_X(4\times 10^{-4}) - V_Y(2\times 10^{-4}) = 30\times 10^{-4}$$

$$-V_X(2\times 10^{-4}) + V_Y(3\times 10^{-4}) = 0$$

$$V_Y = \frac{\begin{vmatrix} 4\times 10^{-4} & 30\times 10^{-4} \\ -2\times 10^{-4} & 0 \end{vmatrix}}{\begin{vmatrix} 4\times 10^{-4} & -2\times 10^{-4} \\ -2\times 10^{-4} & 3\times 10^{-4} \end{vmatrix}} = \frac{60\times 10^{-8}}{8\times 10^{-8}} = 7.5\ volts$$

Replacing the voltage source with a short-circuit and the current source with an open-circuit and calculating $R_{ab} = R_{th}$.

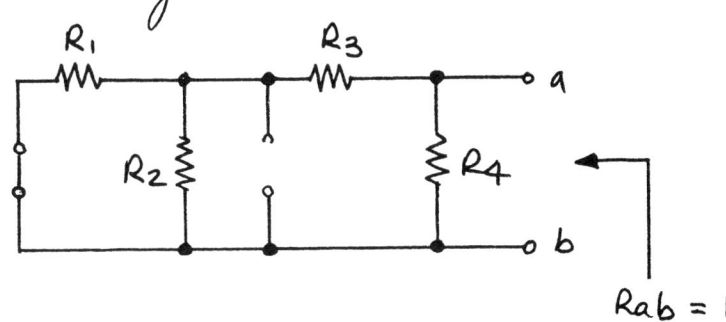

$R_{ab} = R_4 \,\|\, (R_3 + R_1 \,\|\, R_2) = 10K \,\|\, (5K + 10K \,\|\, 10K)$

$R_{ab} = 5\,K\Omega$

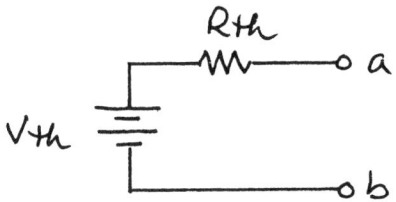

$V_{th} = 7.5 \text{ volts}$

$R_{th} = 5\,K\Omega$

3-35

Use source transformations to find the voltage on the 2Ω resistor.

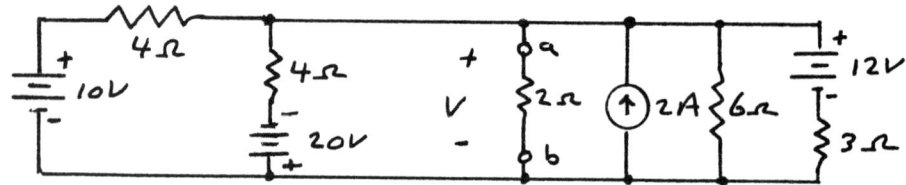

***

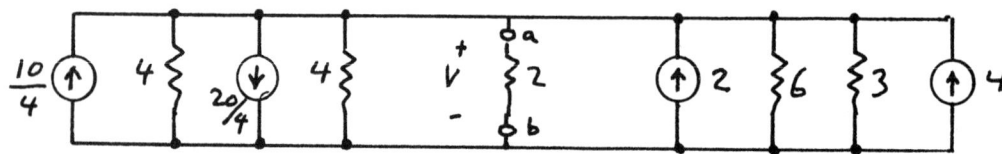

The resistors are all in parallel

$$4\|4\|6\|3 = \frac{1}{1/4 + 1/4 + 1/6 + 1/3} = 1\,\Omega$$

Currents add algebraically

$$I = \frac{10}{4} - \frac{20}{4} + 2 + 4 = 3\tfrac{1}{2}\,A$$

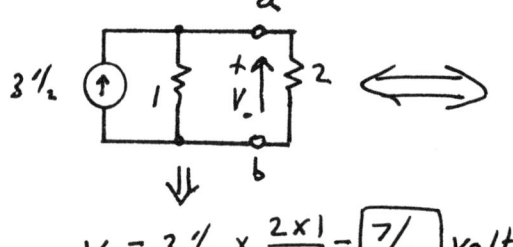

$$V = 3\tfrac{1}{2} \times \frac{2 \times 1}{2+1} = \boxed{7/3}\ \text{volts} \qquad V = 3\tfrac{1}{2} \times \frac{2}{1+2} = \boxed{7/3}\ \text{volts}$$

MAXIMUM POWER TRANSFER

3-36

Find the maximum power to the load R_L in the circuit shown, if the condition for maximum power transfer to R_L is satisfied.

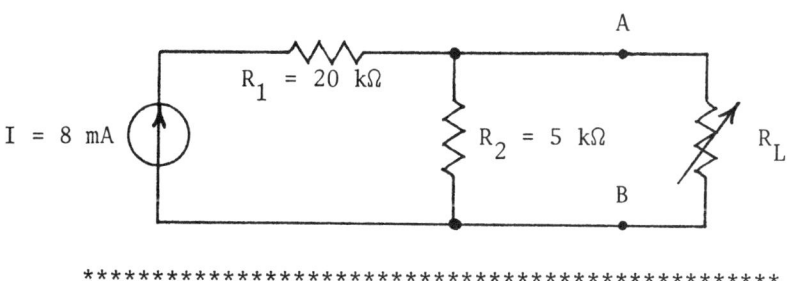

***

1) Maximum Power Transfer occurs when $R_L = R_{Th} = R_N$:
   $R_{Th} = R_N = R_{AB} = R_2$
   $\therefore R_L = R_2 = \underline{5 k\Omega}$.

2) Find the maximum power to R_L
   i) using Thevenin's Theorem:
   $V_{Th} = V_{AB} = V_2$
   $= I R_2 = 8m \cdot 5k = 40 V.$
   $P_L^{Max} = \dfrac{V_{Th}^2}{4 R_{Th}} = \dfrac{(40)^2}{4 \cdot 5k} = \underline{80 mW}.$

   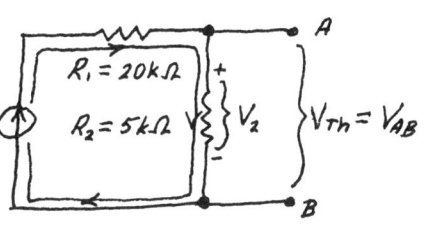

   ii) using Norton's Theorem:
   $I = I_N = 8 mA.$
   $P_L^{Max} = \dfrac{I_N^2 R_N}{4} = \dfrac{(8m)^2 \cdot 5k}{4} = \underline{80 mW}.$

   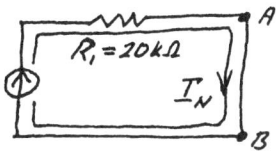

3-37

a) Find the value of R necessary for maximum power transfer.

b) For the R of part a) what power is delivered to the resistor; what power is dissipated by the internal resistors; and what power is supplied by the sources?

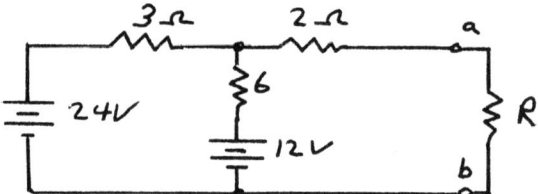

****

a) Find Thevenin circuit looking to the left at terminals a - b

$$R_{eq} = 2 + \frac{3 \times 6}{3+6} = 4\Omega \quad V_T = 24 \times \frac{6}{6+3} + 12 \times \frac{3}{3+6} = 20$$

$\boxed{R = 4\Omega}$ By maximum power transfer theorem

b) From a)

$$I_R = 20/(4+4) = \frac{5}{2} A$$

$$P_R = \left(\frac{5}{2}\right)^2 \times 4 = \boxed{25 W}$$

$$V_1 = I_R(2+4) = 15V, \quad I_1 = \frac{24-15}{3} = 3, \quad I_2 = \frac{15-12}{6} = \frac{1}{2}$$

$$P_{2\Omega} = I_R^2 \times 2 = \left(\frac{5}{2}\right)^2 \times 2 = \boxed{\frac{25}{2} W}$$

$$P_{6\Omega} = I_2^2 \times 6 = \left(\frac{1}{2}\right)^2 \times 6 = \boxed{\frac{3}{2} W}$$

$$P_{3\Omega} = I_1^2 \times 3 = 3^2 \times 3 = \boxed{27 W}$$

$$P_{24V} = -24 I_1 = -24 \times 3 = \boxed{-72 W} \quad \text{supplied by source}$$

$$P_{12V} = 12 I_2 = 12 \times \frac{1}{2} = \boxed{6 W} \quad \text{supplied to source}$$

3-38

a) Find the maximum power that can be obtained from terminals a-b.
b) Repeat (a) for a pure resistive load.

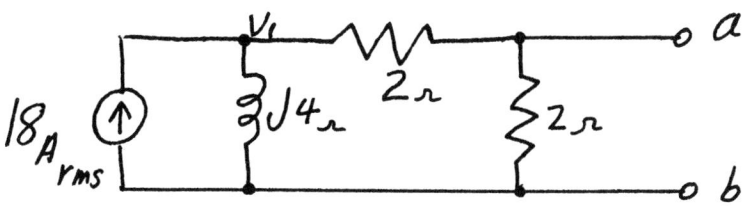

****

Find the Thevenin or Norton equivalent ckt. at a-b
the impedance seen by 18A source is $j4 \| 4$

$$\frac{4(j4)}{4+j4} = \frac{j4}{1+j1} \cdot \frac{1-j1}{1-j1} = \frac{4+j4}{2} = 2+j2$$

$V_1 = 18(2+j2) = 36+j36$

$V_{oc} = \frac{V_1}{2} = 18+j18 = 18\sqrt{2} \angle 45°$

$R_{th} = 2 \| 2+j4 = \frac{2(2+j4)}{4+j4} = \frac{2+j4}{2+j2} \cdot \frac{2-j2}{2-j2} = 1.5+j0.5$

Thevenin Equiv. is $18\sqrt{2}$ V_{rms}, 1.5Ω, $j0.5\Omega$

a) For Max Power Make Load $1.5\Omega - j0.5\Omega$

$$P_{max} = \frac{(V_{rms})^2}{4R_s} = \frac{(18\sqrt{2})^2}{4(1.5)} = \underline{\underline{108 \text{ watts}}}$$

b) For resistive Load make $R_L = |1.5+j0.5| = 1.581\Omega$

$$I_L = \frac{18\sqrt{2}}{1.5+1.581+j0.5} = 8.155 A$$

$$P_L = |I_L|^2 R_L = 8.155^2 \times 1.581 = \underline{\underline{105.156 w}}$$

3-39

Find the maximum power that can be absorbed by a resistor connected across terminals a,b of the following circuit.

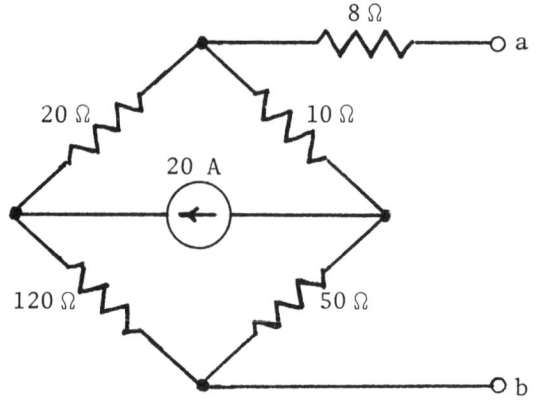

By the maximum power theorem, the resistor that will absorb maximum power has a resistance equal to the Thevenin resistance R_{Th} at terminals a,b, and this power is

$$P_{max} = \frac{0.25 V_{Th}^2}{R_{Th}}$$

in which V_{Th} is the Thevenin or open-circuit voltage at terminals a,b. When we "kill" the current source to find R_{Th}, the 20-Ω and 120-Ω resistors become in series, as do the 10-Ω and 50-Ω resistors, and these two series combinations are in parallel. Consequently,

$$R_{Th} = 8 + (20+120)||(10+50) = 8 + \frac{140(60)}{140+60} = 50 \, \Omega$$

To find the open-circuit voltage, observe that with an open circuit at terminals a,b, no current flows through the 8-Ω resistor, which means that the voltage across it is zero. Consequently, the open-circuit voltage is equal to the sum of the voltage drops, top to bottom, across the 10-Ω and 50-Ω resistors. By current division, the current flowing down through the 10-Ω resistor is

$$\frac{120+50}{120+50+20+10} \, 20 = \frac{170}{200} \, 20 = 17 \, A$$

which produces a voltage drop of $10(17) = 170$ V. The current flowing up through the 50-Ω resistor is $20 - 17 = 3A$, which produces a voltage drop of $-3(50) = -150$ V. So, the Thevenin or open circuit voltage is $170 - 150 = 20$ V, and the maximum power that a resistor can absorb is

$$P_{max} = \frac{0.25 \, V_{Th}^2}{R_{Th}} = \frac{0.25(20)^2}{50} = 2 \, W$$

3-40

Determine the value of R_1 which results in the maximum power dissipated in the load resistor R_L.

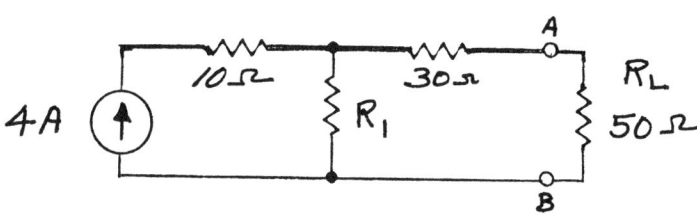

****

For maximum power transfer, the internal resistance looking into terminals A and B must be equal to the load resistance R_L. Since the 4A source has infinite internal resistance, $R_i = 30 + R_1 = R_L = 50$

For maximum power, $R_1 = 50 - 30 = 20 \, \Omega$

3-41

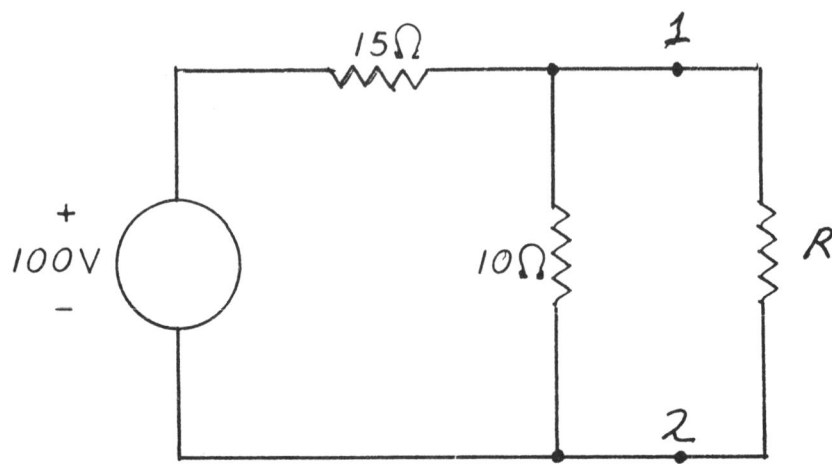

In this circuit find R for maximum power transfer and calculate that power. Also calculate total power lost in internal resistors.

***

First obtain Thevenin equivalent of circuit the left of terminals 1 and 2. With the voltage source dead (short circuit), the 10 ohm and 15 ohm resistors are in parallel.

$$R_{TH} = \frac{15 \times 10}{15+10} = 6 \text{ ohms.}$$

The open circuit voltage at 1-2 is obtained by voltage division.

$$V_{OC} = \frac{10}{10+15} \times 100 = 40 \text{ volts}$$

For maximum power transfer R equals $R_{TH} = 6$ ohms. The current through R is then $\frac{40}{6+6} = \frac{10}{3}$ A and the maximum power is:

$$P = I^2 R = \left(\frac{10}{3}\right)^2 \times 6 = 66.667 \text{ watts.}$$

The voltage across the 10 ohm resistor is the same as that across R.
$$V_{12} = \frac{10}{3} \times 6 = 20 \text{ volts}$$

The current through the 10 ohm resistor is $\frac{20}{10} = 2$ Amps, and the power dissipated in it is
$$P_{10} = 2^2 \times 10 = 40 \text{ watts}$$

By the current law at node 1 the current from the source through the 15 ohm resistor is
$$I_{(source)} = I_{15} = \frac{10}{3} + 2 = \frac{16}{3} \text{ Amps.}$$

Power dissipated in the 15 ohm resistor $P_{15} = \left(\frac{16}{3}\right)^2 \times 15 = 426.667$ watts

Check:
$$P_{source} = 100 \times \frac{16}{3} = \underline{\underline{533.333}} \text{ watts}$$

Power dissipated $= P_R + P_{10} + P_{15}$
$$= 66.6667 + 40 + 426.667 = \underline{\underline{533.333}} \text{ W}$$

3-42

If, in the circuit shown, we can vary R and C,

A. What values of R and C draw maximum power from the generator, at 1 MHz?
B. What is the Q of the resulting resonance?

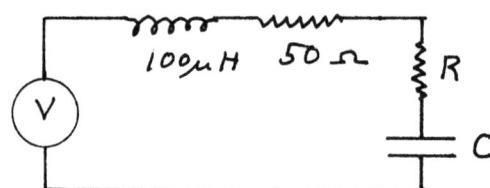

****

A. We draw maximum power out of a generator when we cause it to drive the complex conjugate of its internal impedance, which here includes the inductor and 50 Ω resistor. Thus we require:

$\boxed{R = 50\,\Omega}$ and $\dfrac{1}{\omega C} = \omega L$; $C = \dfrac{1}{\omega^2 L}$

$= \dfrac{1}{(2\pi \times 10^6)^2 \, 100 \times 10^{-6}} = \dfrac{10^4}{4\pi^2 \, 10^8} = \boxed{2.53\,\mu F}$

B. $Q = \dfrac{\omega L}{R_{total}} = \dfrac{2\pi \times 10^{+6} \times 100 \times 10^{-6}}{50 + 50} = 2\pi = \boxed{6.28}$

4
NETWORK GRAPHS

THE GRAPH OF A CIRCUIT

━━━━━━━━━━━━━━━━━━━━━━━━━━━━━━ 4-1

Draw the directed Linear graph of the network shown and write the incident matrix for the graph.

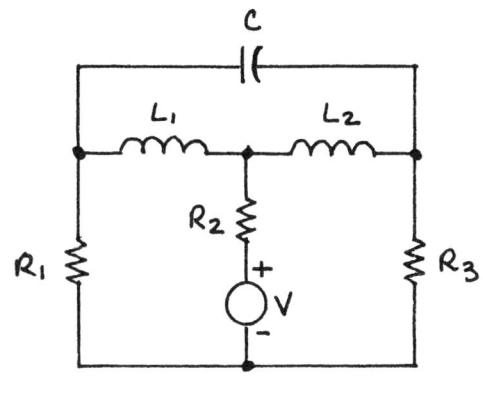

****

Replacing each element with a straight line segment and identifying each node

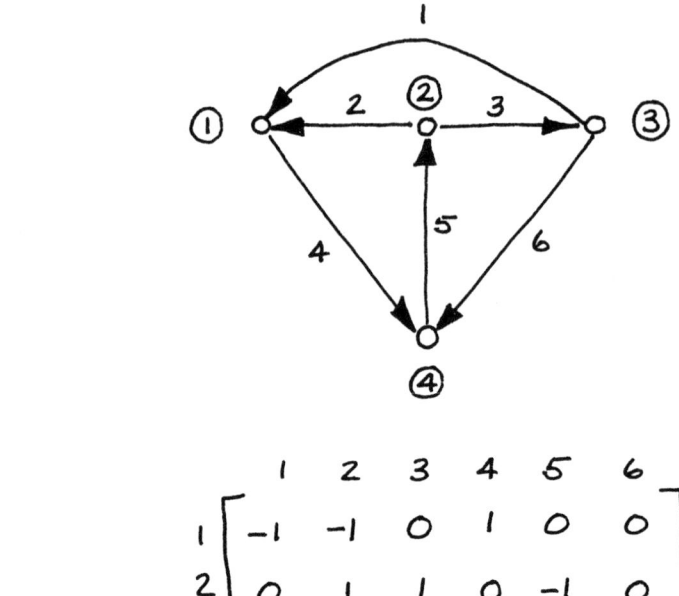

$$A_I = \begin{array}{c} \\ 1 \\ 2 \\ 3 \\ 4 \end{array} \begin{array}{c} \begin{array}{cccccc} 1 & 2 & 3 & 4 & 5 & 6 \end{array} \\ \left[\begin{array}{cccccc} -1 & -1 & 0 & 1 & 0 & 0 \\ 0 & 1 & 1 & 0 & -1 & 0 \\ 1 & 0 & -1 & 0 & 0 & 1 \\ 0 & 0 & 0 & -1 & 1 & -1 \end{array} \right] \end{array}$$

4-2

Find the total number of trees in the <u>dual</u> of the graph shown.

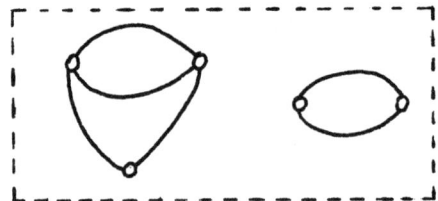

* *

FIRST, FIND DUAL GRAPH:

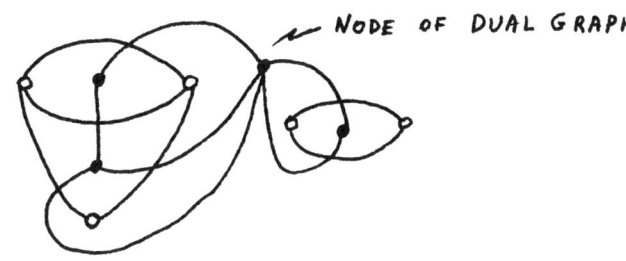
← NODE OF DUAL GRAPH

DUAL GRAPH =

ASSIGN NODE D AS REFERENCE NODE. THEN THE TOTAL NUMBER OF TREES IS GIVEN BY DET$\{T\}$. THE ELEMENTS OF T ARE:

t_{jj} = NUMBER OF BRANCHES INCIDENT AT NODE j.
t_{ij} = $-$[NUMBER OF BRANCHES HANGING BETWEEN NODES i AND j]

$$T = \begin{array}{c|ccc} & A & B & C \\ \hline A & 2 & -1 & 0 \\ B & -1 & 3 & 0 \\ C & 0 & 0 & 2 \end{array}$$

DET$\{T\}$ = 2(6) + 1(−2)
= 10 TREES

TREES, COTREES, AND CUTSETS

4-3 ■■

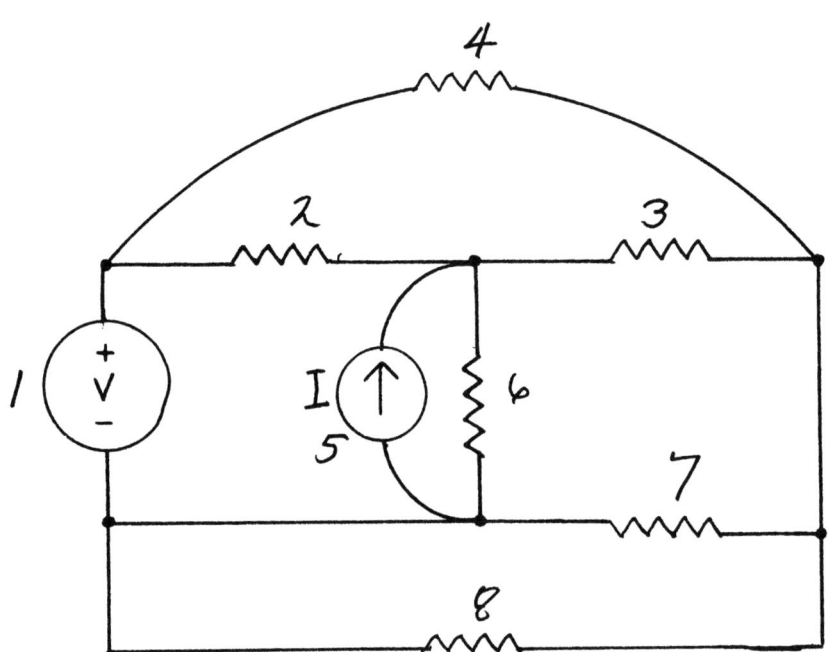

Draw a topological graph of the network, indicate a suitable tree containing all voltage sources and no current sources. State:

  Number of branches _____
  Number of nodes _____
  Number of tree branches _____
  Number of link branches _____

****

Trees, Cotrees, And Cutsets / 127

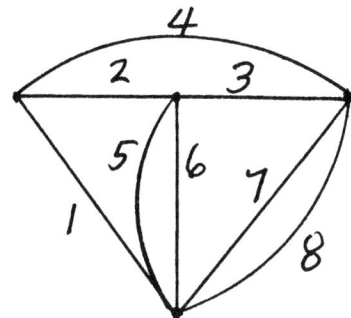

Branches — 8
Nodes — 4
Tree branches — 3
Link branches — 5

There are a number of possible trees. Three are given here.

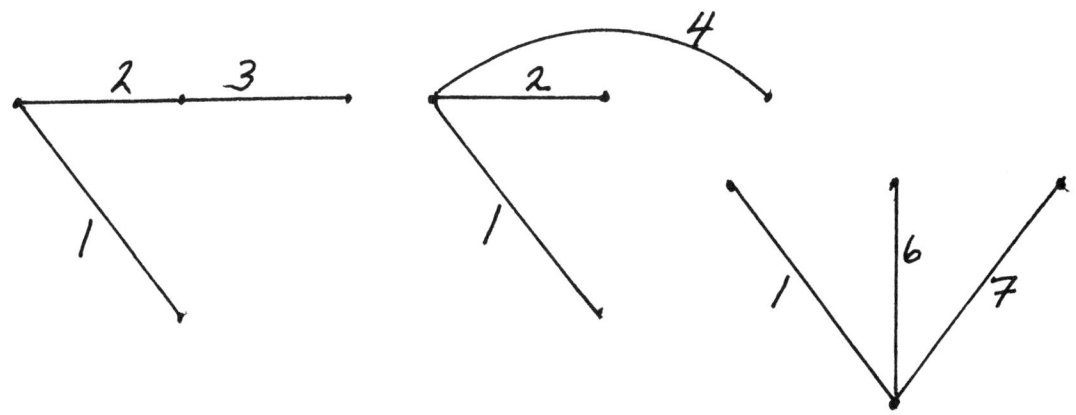

4-4

For the circuit shown draw a tree and the corresponding co-tree. Solve for the voltage v_x. What is the voltage across the current source? Determine the currents shown.

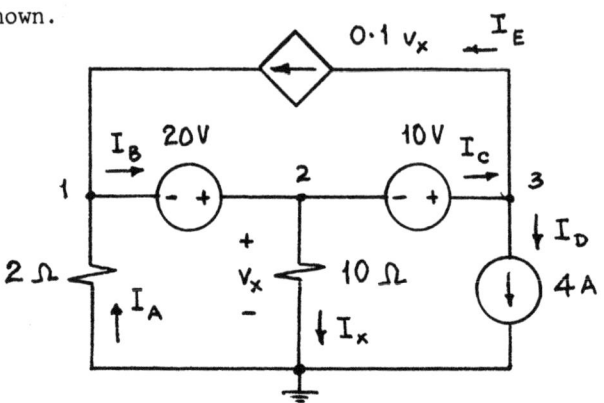

****

A tree is shown in firm lines while the co-tree is shown in broken lines.

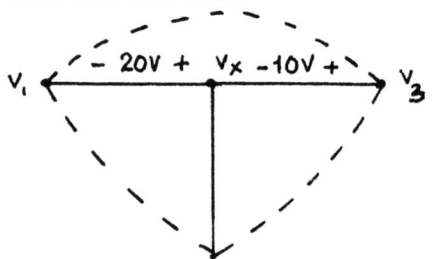

Identifying the grounded node as the reference node, the node voltages V_1, V_x and V_3 are shown. Since the tree branch voltages between V_1 and V_x and also between V_x and V_3 are known, the three non-reference nodes are taken as one "supernode".

Writing the KCL eqn. for the supernode:
$$\frac{V_1}{2} + \frac{V_x}{10} + 4 = 0.$$

Note that the dependent current source $0.1\, v_x$ does not appear in the eqn. because it is contained within the supernode. Since $V_1 = -20 + V_x$, substitution in the KCL eqn. gives,

$$\frac{-20 + V_x}{2} + \frac{V_x}{10} + 4 = 0.$$

Solving: $V_x = 10$ V

The current source voltage = $v_3 = 10 + v_x$
$\quad = 10 + 10$ V $= 20$ V

$I_x = \dfrac{v_x}{10} = \dfrac{10}{10} = 1$ A. ; $\quad I_D = 4$ A.

$I_A = I_x + I_D = 1$ A $+ 4$ A $= 5$ A.

$I_E = 0.1\, v_x = 0.1 \times 10$ A $= 1$ A.

$I_B = I_E + I_A = 1$ A $+ 5$ A $= 6$ A.

$I_C = I_B - I_x = 6$ A $- 1$ A $= 5$ A.

4-5

For the network shown determine the number of:
1. Node-pairs.
2. Branch trees.
3. Node trees.

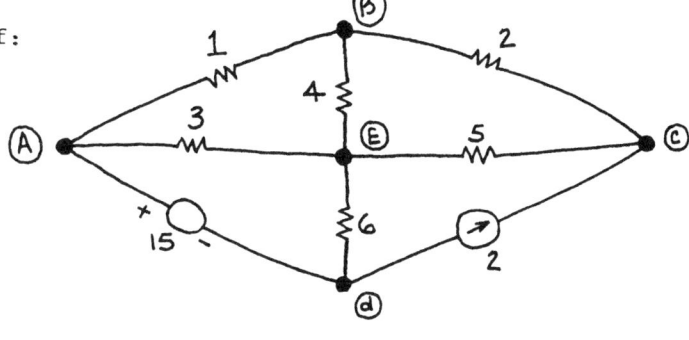

***

Number of node-pairs $= \dfrac{5!}{2!\, 3!} = 10$

Number of node trees $= n^{n-2} = 5^3 = 125$

Number of branch trees $=$ det of the matrix

$$\begin{array}{c} \\ A \\ B \\ C \\ E \end{array} \begin{array}{cccc} A & B & C & E \end{array}$$
$$\begin{bmatrix} 3 & -1 & 0 & -1 \\ -1 & 3 & -1 & -1 \\ 0 & -1 & 3 & -1 \\ -1 & -1 & -1 & 4 \end{bmatrix} = M_T \qquad \text{DET } M_T = 45$$

INDEPENDENT VOLTAGE EQUATIONS AND FUNDAMENTAL CUTSETS

4-6

a.) For the circuit shown below construct a graph of the circuit and on this graph determine a tree.

b.) Using generalized node analysis, determine v_1.

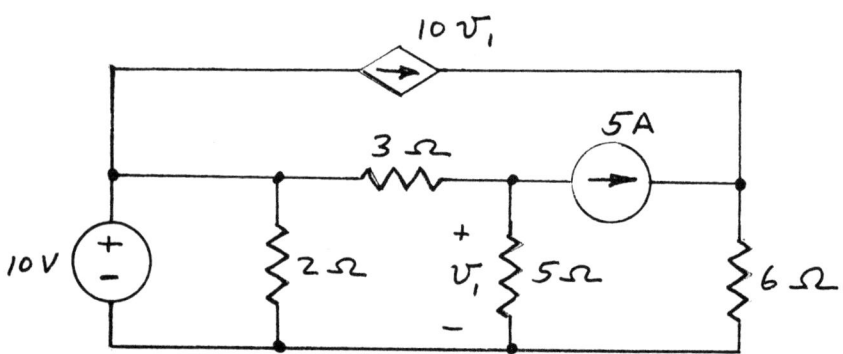

****

a)

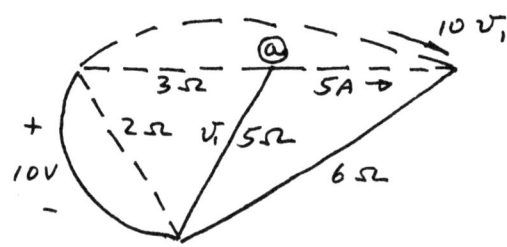

b.) KCL NODE @ :

$$0 = 5 + \frac{V_1}{5} + \frac{V_1 - 10}{3}$$

$$V_1 = -3.125 \text{ VOLTS} \qquad \text{ANS}$$

INDEPENDENT CURRENT EQUATIONS AND FUNDAMENTAL LOOPS

4-7

For the oriented graph below, and the particular tree indicated, obtain the fundamental loop matrix B_f. Partition B_f in the form:

$$B_f = \left[\, U_3 \mid H \,\right]$$

and from the relation $C_f B_f^{\,t} = 0$, obtain the fundamental cut-set matrix for the case of the chosen tree.

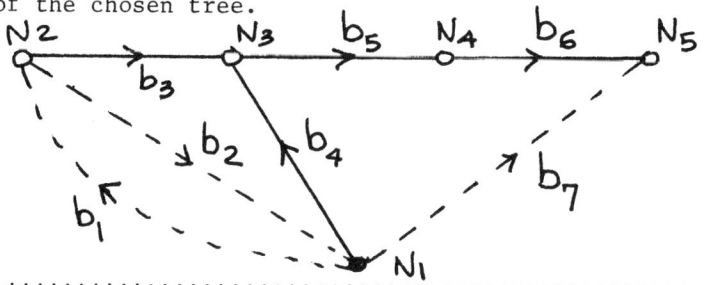

***

The fundamental loop matrix is

$$B_f = \begin{bmatrix} b_1 & b_2 & b_3 & b_4 & b_5 & b_6 & b_7 \\ 1 & 0 & 1 & -1 & 0 & 0 & 0 \\ 0 & 1 & -1 & 1 & 0 & 0 & 0 \\ 0 & 0 & 0 & -1 & -1 & -1 & 1 \end{bmatrix}$$

Moving the column b_7 next to b_2, we can partition B_f as shown below:

$$B_f = \begin{bmatrix} b_1 & b_2 & b_7 & b_3 & b_4 & b_5 & b_6 \\ 1 & 0 & 0 & 1 & -1 & 0 & 0 \\ 0 & 1 & 0 & -1 & 1 & 0 & 0 \\ 0 & 0 & 1 & 0 & -1 & -1 & -1 \end{bmatrix} = \left[\, U_3 \mid H \,\right]$$

If the corresponding fundamental cutset matrix is partitioned as $C_f = [G \mid U_4]$, then the relation

$$C_f B_f^t = 0 \quad \text{yields}$$

$$G + H^t = 0 \quad \text{or} \quad G = -H^t.$$

Since

$$H = \begin{bmatrix} 1 & -1 & 0 & 0 \\ -1 & 1 & 0 & 0 \\ 0 & -1 & -1 & -1 \end{bmatrix},$$

we have

$$G = \begin{bmatrix} -1 & 1 & 0 \\ 1 & -1 & 1 \\ 0 & 0 & 1 \\ 0 & 0 & 1 \end{bmatrix}$$

The complete fundamental cut-set matrix is

$$C_f = \begin{array}{c} b_1\ b_2\ \ b_7\ b_3\ b_4\ b_5\ b_6 \\ \begin{bmatrix} -1 & 1 & 0 & 1 & 0 & 0 & 0 \\ 1 & -1 & 1 & 0 & 1 & 0 & 0 \\ 0 & 0 & 1 & 0 & 0 & 1 & 0 \\ 0 & 0 & 1 & 0 & 0 & 0 & 1 \end{bmatrix} \end{array}$$

5
TRANSIENT ANALYSIS

INDUSTORS AND CAPACITORS

━━━ 5-1

At t = 0, a current generator which generates the current waveform shown is connected across an initially uncharged capacitor C. Determine the voltage waveform across the capacitor from t = 0 to t = 2.5 seconds and sketch this waveform.

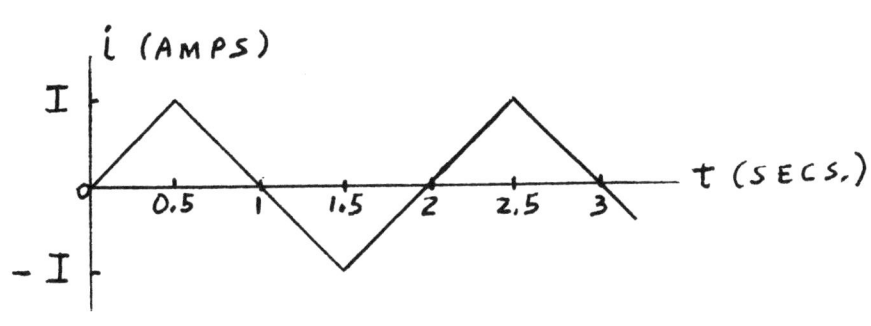

* *

TIME INTERVAL (SECONDS)	CURRENT (AMPS.)
$0 \leq t \leq 0.5$	$i = 2It$
$0.5 \leq t \leq 1.5$	$i = -2It + 2I$
$1.5 \leq t \leq 2.5$	$i = 2It - 4I$

$$v = \frac{1}{C}\int_0^t i\, dt$$

For $0 \le t \le 0.5$, $\quad v = \frac{1}{C}\int_0^t 2It\, dt = \frac{I}{C}t^2$

At $t = 0.5$ secs., $v = \frac{I}{4C}$ Volts.

For $0.5 \le t \le 1.5$, $\quad v = \frac{1}{C}\int_0^{0.5} i\, dt + \frac{1}{C}\int_{0.5}^t (-2It + 2I)\, dt$

$v = \frac{I}{4C} + \frac{I}{C}\left[-t^2 + 2t - \frac{3}{4}\right] = \frac{I}{C}\left[-t^2 + 2t - \frac{1}{2}\right]$

At $t = 1.5$ secs., $v = \frac{I}{4C}$ Volts

For $1.5 \le t \le 2.5$, $\quad v = \frac{1}{C}\int_0^{1.5} i\, dt + \frac{1}{C}\int_{1.5}^t (2It - 4I)\, dt$

$v = \frac{I}{4C} + \frac{I}{C}\left[t^2 - 4t + 3.75\right]$

$v = \frac{I}{C}\left[t^2 - 4t + 4\right]$ Volts

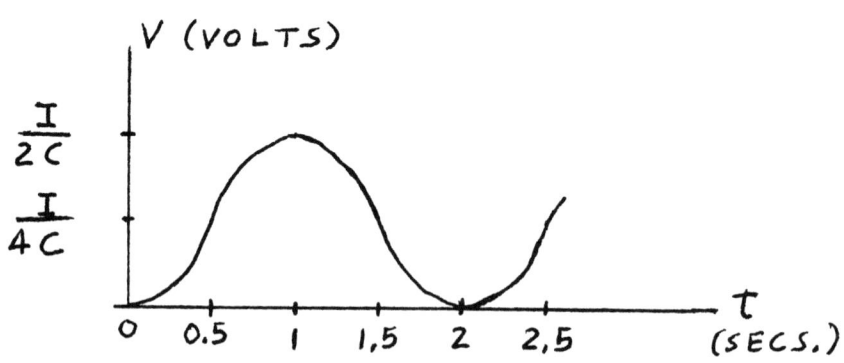

5-2

A coil has an inductance of 1 Henry and negligible resistance. It carries the current shown.

(i) Find the maximum energy stored in the inductor.
(ii) Find the maximum voltage across the inductor.
(iii) Plot the coil voltage against time.
(iv) Now assume that the resistance of the coil is 10 ohms. What is the energy lost due to the passage of the current in the coil?

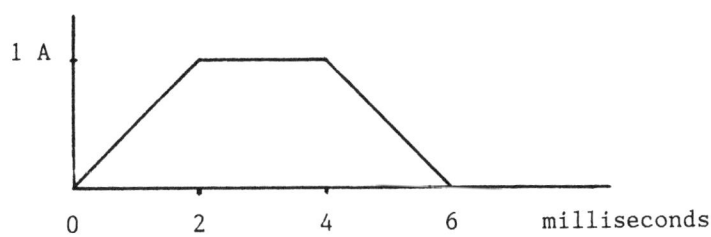

****

(i) Max. energy stored = $\frac{1}{2} L i_{max}^2 = \frac{1}{2} \times 1 \times 1^2 = \frac{1}{2}$ J.

(ii) $V_{coil\ max} = L\left(\frac{di}{dt}\right)_{max} = 1 \times \frac{1}{2 \times 10^{-3}} = 0.5 \times 10^3 = 500$ V.

(iii) The coil voltage will be negative between 4 ms and 6 ms because the current is decreasing.

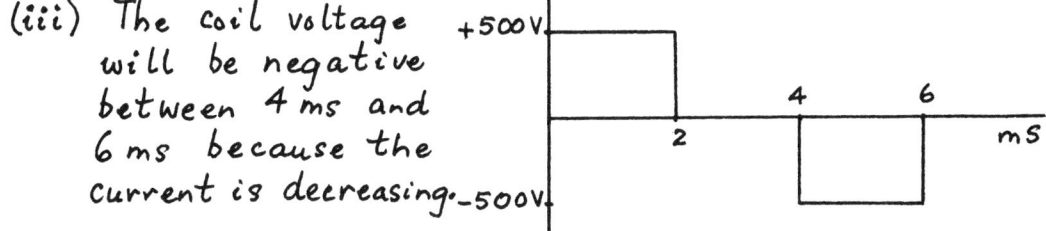

(iv) Power $p = i^2 R$. Energy $= \int p\,dt = \int i^2 R\,dt$. The p versus t plot is shown. The energy is the area under the curve. Noting that the area for 0 – 2 ms and 4 – 6 ms are the same, the energy is $2\int_0^{2\times 10^{-3}} \left(\frac{t}{2\times 10^{-3}}\right)^2 \times 10\,dt + 1^2 \times 10 \times 2 \times 10^{-3}$ J.

$= \frac{20}{4 \times 10^{-6}} \int_0^{2 \times 10^{-3}} t^2 dt + 20 \times 10^{-3}$ J $= \frac{5}{10^{-6}}\left(\frac{t^3}{3}\right)_0^{2\times 10^{-3}} + 20\times 10^{-3} = 33.33 \times 10^{-3}$ J.

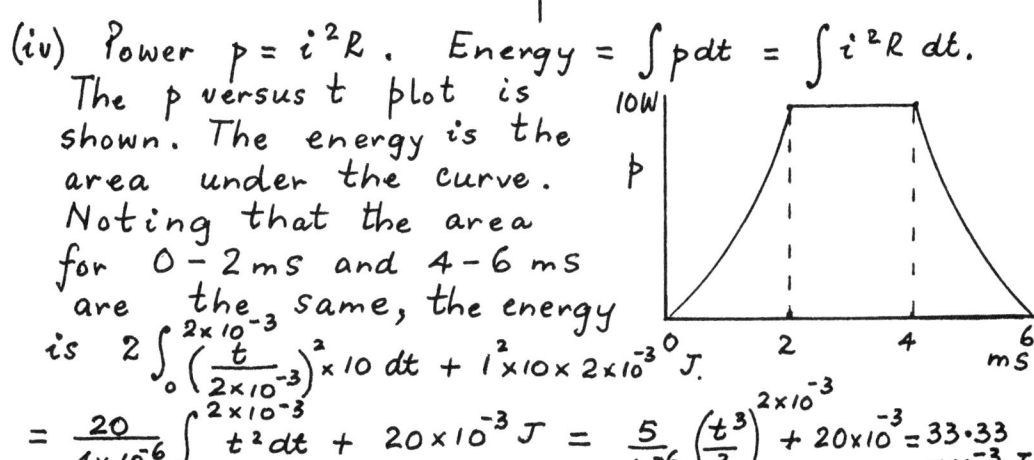

5-3

Determine the value of R such that the energies stored in the inductor and capacitor are equal in the dc steady state.

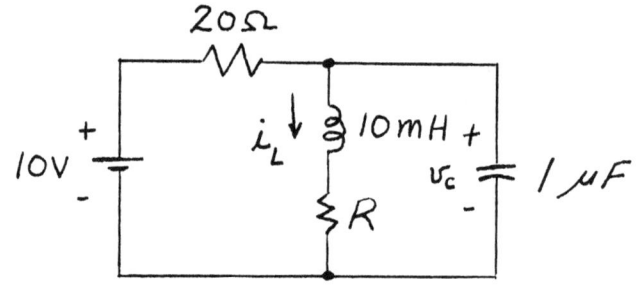

***

WE WANT $\quad \frac{1}{2} L i_L^2 = \frac{1}{2} C v_c^2$

BUT, IN THE DC STEADY STATE, $i_L = v_c/R$.

THEREFORE, $\quad L\left(\dfrac{v_c}{R}\right)^2 = C v_c^2$

OR $\quad \dfrac{L}{R^2} = C$

AND $\quad R = \sqrt{L/C} = \sqrt{10^4}$

OR $\quad R = 100 \, \Omega$

5-4

For the boundary conditions given below, calculate the two exponential term coefficients for the capacitor voltage expression.

$i_c(0) = 50$ mA.
$v_c(t) = 5$ V.; $t \leq 0$
$v_c(t) = A_1 t e^{-100t} + A_2 e^{-50t}$ V.; $t \geq 0$

****

$v_c(0) = 5 = A_1 t e^0 + A_2 e^0$

$\Rightarrow \underline{A_2 = 5}$

$i_c(t) = C \dfrac{dv_c}{dt} = (100 \times 10^{-6})[A_1 e^{-100t} - A_1(100) t e^{-100t} - A_2(50) e^{-50t}]$

$i_c(0) = 50 \times 10^{-3} = 10^{-4}[A_1 e^0 - (5)(50) e^0]$

$\Rightarrow \underline{A_1 = 750}$

5-5

Calculate VA(t).

$\{t e^{-2t} \sin 3t\} u(t)$

****

$VA(t) = L \dfrac{d\{I_{source}\}}{dt}$

$= 2\left[e^{-2t} \sin 3t - 2t e^{-2t} \sin 3t + 3t e^{-2t} \cos 3t\right] u(t)$

5-6

The switch is opened at t=0. Determine $i_L(0^+)$, $v_L(0^+)$ and $\frac{dv_L(0^+)}{dt}$.

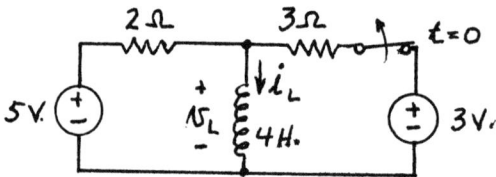

****

$i_L(0^-) = \frac{5}{2} + \frac{3}{3} = \frac{7}{2}$ A.   BUT, $i_L(0^-) = i_L(0^+) = \frac{7}{2}$ A.

$v_L(0^+) = 5 - 2 i_L(0^+) = 5 - 2\left(\frac{7}{2}\right) = -2$ V.

WITH SWITCH OPEN, $5 = 2 i_L + 4 \frac{di_L}{dt}$ (KVL)

DIFFERENTIATING, $0 = 2 \frac{di_L}{dt} + \frac{dv_L}{dt}$

$\therefore \frac{dv_L}{dt}(0^+) = -2 \frac{di_L}{dt}(0^+) = -2 \left(\frac{v_L(0^+)}{4} \right)$

$\frac{dv_L}{dt}(0^+) = 1$ V/SEC

5-7

Given the following current in a 25 mH inductor, sketch the voltage across the inductor.

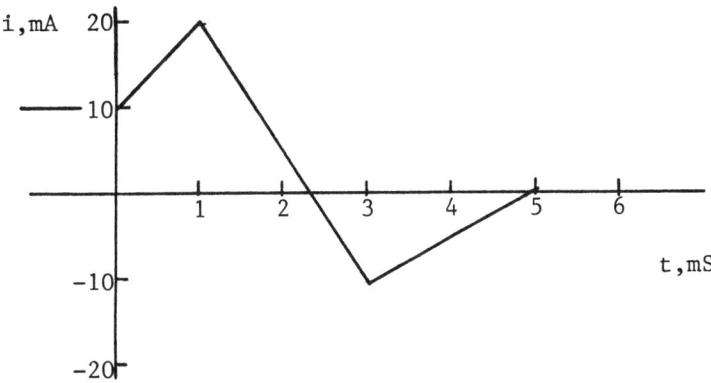

****

For Piecewise Linear Segments, The Voltage Across The Inductor Can be Expressed As

$$v_L(t) = L \, di/dt = L \cdot \text{Slope} = L \, \Delta I/\Delta t$$

For $t < 0$ $\quad v_L(t) = 25 \times 10^{-3} \cdot d(10mA)/dt \equiv 0 \, V.$

For $0 < t < 1$ $\quad v_L(t) = 25 \times 10^{-3} \, \dfrac{(20-10)}{(1-0)} \, \dfrac{mA}{mS} = 25 \times 10^{-3} \times 10 = 250 \, mV.$

For $1 < t < 3$ $\quad v_L(t) = 25 \times 10^{-3} \, \dfrac{(-10-20)}{(3-1)} \, \dfrac{mA}{mS} = 25 \times 10^{-3} \times \left(\dfrac{-30}{2}\right) = -375 \, mV.$

For $3 < t < 5$ $\quad v_L(t) = 25 \times 10^{-3} \, \dfrac{(0-(-10))}{(5-3)} \, \dfrac{mA}{mS} = 25 \times 10^{-3} \times \dfrac{10}{2} = 125 \, mV.$

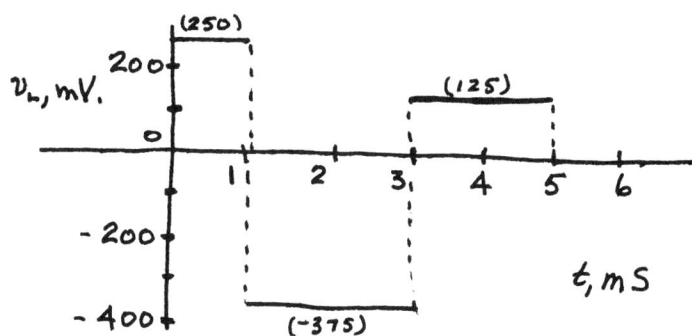

5-8

The current flowing into a ½ F capacitor is 16 cos 4t amperes and the voltage on the capacitor is zero at t=zero. Find the energy stored in the capacitor at π/8 seconds and the power delivered to the capacitor at π/12 seconds.

***

First find the voltage on the capacitor.

$$V = \frac{1}{C}\int_0^t I\,dt = \frac{1}{\frac{1}{2}}\int_0^t 16\cos 4t\,dt$$

$$V = \frac{2 \times 16}{4}\sin 4t = 8\sin 4t \text{ Volts}$$

At $\frac{\pi}{8}$ seconds $\sin 4t = \sin\frac{\pi}{2} = 1$ and $V = 8$ volts. Energy stored $W = \frac{1}{2}CV^2$

$$W = \frac{1}{2} \times \frac{1}{2} \times 8^2 = \underline{16 \text{ Joules.}}$$

At $\frac{\pi}{12}$ seconds $\sin 4t = \sin\frac{\pi}{3} = \frac{\sqrt{3}}{2}$

and $\cos 4t = \cos\frac{\pi}{3} = \frac{1}{2}$

$$I = 16 \times \frac{1}{2} = 8 \text{ Amps.}$$

$$V = 8 \times \frac{\sqrt{3}}{2} = 4\sqrt{3} \text{ Volts}$$

Power $P = VI = 8 \times 4\sqrt{3} = \underline{55.426 \text{ Watts.}}$

5-9

The switch is opened at t=0. (a) Find I at t=0⁻.
(b) Find the voltage across the open switch at t=0⁺.

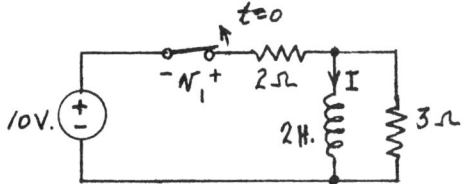

*****

a. $I = \frac{10}{2} = 5A.$ (INDUCTOR IS A SHORT), $I = i(0^-) = 5A.$

b. KVL: $v_1(0^+) + 10 = 3(-5)$

$\therefore v_1(0^+) = -25 V.$

L LOOKS OPEN, $i(0^+) = 5A.$

5-10

Given the initial current and the applied voltage, first determine the inductor time domain current expression and then determine how much energy is stored in the magnetic field at 2 milliseconds.

$i_L(0) = 0.75$ A.
$v_L(t) = 10e^{-3t}$ V.; $t \geq 0$

*****

$i_L(t) = i_L(0) + \frac{1}{L}\int_0^t v_L(\tau)d\tau = 0.75 + \frac{1}{0.05}\int_0^t 10e^{-3\tau}d\tau$

$i_L(t) = 67.42 - 66.67e^{-3t}$ A.; $t \geq 0$

$i_L(0.002) = 1.15 A.$ & $w_L(t) = \frac{1}{2}L[i_L(t)]^2$

$\underline{w_L(0.002) = 0.33 \text{ J.}}$

5-11

For each of the circuits determine the capacitor voltages after the capacitors have charged to their final voltages.

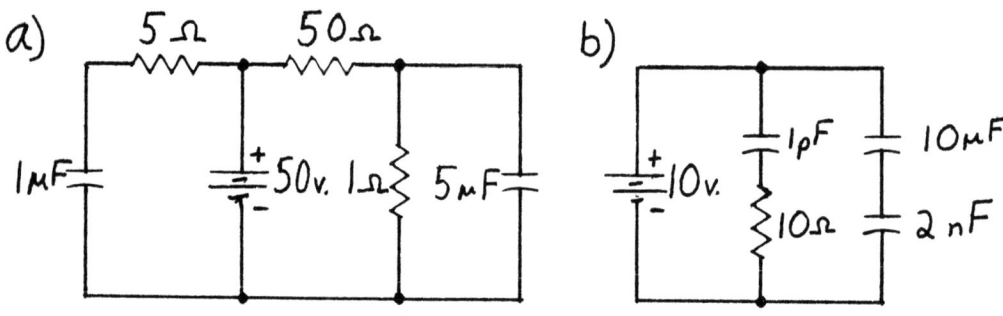

****

a) $V_{1\mu F} = \underline{\underline{50 \text{ VOLTS}}}$

$V_{5\mu F} = 50 \cdot \dfrac{1}{1+50} = \underline{\underline{.980 \text{ VOLTS}}}$

b) $V_{1pF} = \underline{\underline{10 \text{ VOLTS}}}$

TOTAL CAPACITANCE OF THE $10\mu F$ AND $2nF$ BRANCH $= \left(\dfrac{1}{10 \times 10^{-6}} + \dfrac{1}{2 \times 10^{-9}} \right)^{-1}$

$\cong 2.00 \times 10^{-9} F$

TOTAL Q IN THAT BRANCH $\cong CV = (2.00 \times 10^{-9})(10)$

$\cong 2.00 \times 10^{-8}$ COULOMBS

$V_{10\mu F} = Q/10\mu F$
$= 2.00 \times 10^{-8} / 10 \times 10^{-6} \cong \underline{\underline{2.00 \times 10^{-3} \text{ VOLTS}}}$

$V_{2nF} = Q/2nF$
$= 2.00 \times 10^{-8} / 2 \times 10^{-9} \cong \underline{\underline{10.0 \text{ VOLTS}}}$

SINGLE TIME CONSTANT CIRCUITS

5-12

Find V_C as a function of time and plot it versus time if the switch is closed at t=0. What is V_C at t=.5 seconds (to 6 figures).

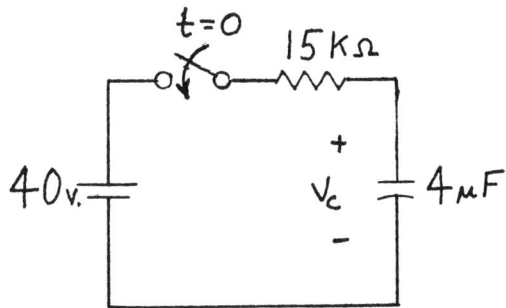

****

$$V_C = A + Be^{-t/\tau} \quad A = V_C(\infty) = 40V \quad B = V_C(0+) - V_C(\infty) = -40$$

$$V_C = V_S(1 - e^{-t/\tau}) \quad \tau = RC = (15 \times 10^3)(4 \times 10^{-6})$$
$$= .06 \text{ SECONDS}$$

$$\underline{\underline{V_C = 40(1 - e^{-t/.06})}}$$

t	V_C
0	0
.1	32.4
.2	38.6
.3	39.7
.5	39.9904

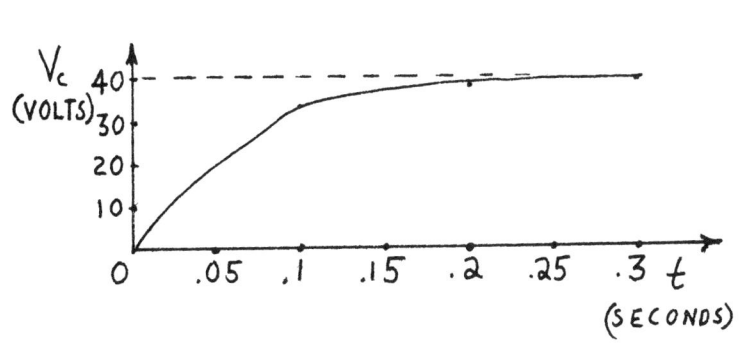

at t=.5s $\underline{V_C = 39.9904 \text{ VOLTS}}$

5-13

Find i(t) for t > 0.

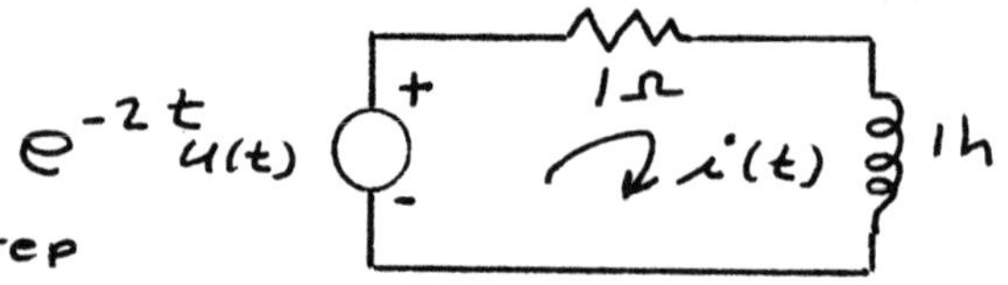

Note: u(t) = unit step

Source: $e^{-2t} u(t)$, resistor $1\,\Omega$, inductor $1\,H$, current $i(t)$

***

For $t < 0$, $u(t) = 0 \rightarrow i(t) = 0$

For $t > 0$

$$1\frac{di}{dt} + 1\cdot i = e^{-2t}$$

First, find the forced solution

try $i_p^\circ = Ae^{-2t}$

$$-2Ae^{-2t} + Ae^{-2t} = e^{-2t} \longrightarrow A = -1$$

Next, find the homogeneous solution

$$\frac{di_h}{dt} + i_h = 0 \rightarrow i_h = Be^{-t}$$

Total solution

$$i_T = -e^{-2t} + Be^{-t}$$

$$i_T(0) = 0 = -1 + B$$

$$\boxed{i = -e^{-2t} + e^{-t}}$$

5-14

For the circuit shown determine: $i_1(0^+)$, $i_2(0^+)$, energy stored in L_1 and L_2 at $t = 0$, $i_1(t)$ and $i_2(t)$ for $t>0$.

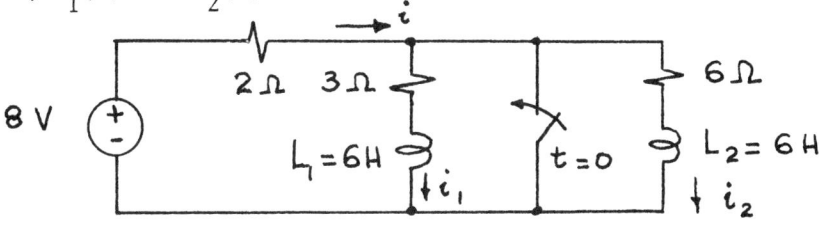

The (dc steady state) circuit for $t<0$ is shown

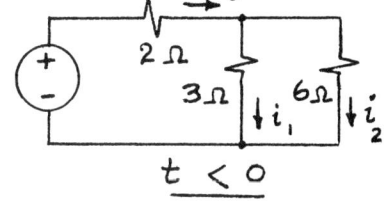

So $i(0^-) = \dfrac{8}{2 + \dfrac{3 \times 6}{3+6}} = 2$ A

By the current divider rule, $i_1(0^-) = 2 \times \dfrac{6}{9} = \dfrac{4}{3}$ A

and $i_2(0^-) = 2 \times \dfrac{3}{9} = \dfrac{2}{3}$ A

Since currents in inductors cannot change instantaneously, $i_1(0^-) = i_1(0^+) = \dfrac{4}{3}$ A.

and $i_2(0^-) = i_2(0^+) = \dfrac{2}{3}$ A.

Energy stored in L_1 at $t=0$ is,

$\dfrac{1}{2} L_1 i_1^2(0^+) = \dfrac{1}{2} \times 6 \times \left(\dfrac{4}{3}\right)^2 = \dfrac{16}{3}$ J.

Similarly, energy stored in L_2 at $t=0$ is,

$\dfrac{1}{2} L_2 i_2^2(0^+) = \dfrac{1}{2} \times 6 \times \left(\dfrac{2}{3}\right)^2 = \dfrac{4}{3}$ J.

At $t=0$ the switch is closed. The stored energies in L_1 and L_2 flow through the closed switch and get dissipated in the 3Ω and 6Ω resistances.

For $t>0$: $i_1(t) = i_1(0^+) e^{-\frac{3\Omega}{6H} \cdot t} = \dfrac{4}{3} e^{-\frac{t}{2}}$ A.

and $i_2(t) = i_2(0^+) e^{-\frac{6\Omega}{6H} t} = \dfrac{2}{3} e^{-t}$ A.

5-15

In the network below, the switch has been in position A long enough for all transients to have died out. Sketch the response versus time for 5 time constants of the voltage across R_3 after the switch is moved to position B.

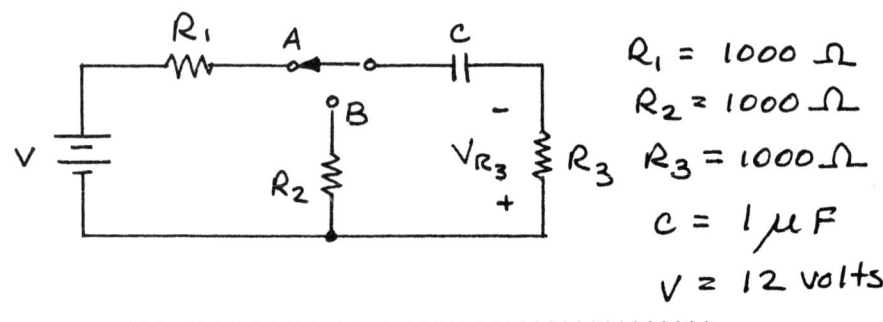

$R_1 = 1000 \, \Omega$
$R_2 = 1000 \, \Omega$
$R_3 = 1000 \, \Omega$
$C = 1 \, \mu F$
$V = 12 \text{ volts}$

***

The voltage across the capacitor is 12 volts. When the switch is moved to position B, the circuit becomes

Time Constant is $\tau = (R_2 + R_3)C$

$= (2000)(1\mu)$

$\tau = 2 \text{ m sec's}$

$$0 = i(t)(R_2 + R_3) + \left[\frac{1}{C}\int_0^t i(t)dt + V_0\right]$$

$$0 = (R_2 + R_3)\frac{di(t)}{dt} + \frac{1}{C}i(t)$$

$$i(t) = I_0 \, e^{\frac{-t}{(R_2+R_3)C}} \qquad I_0 = \frac{V_0}{R_2+R_3} = \frac{12}{2000} = 6 \times 10^{-3}$$

$$i(t) = 6 \times 10^{-3} e^{-500t}$$

$$\boxed{v_{R_3}(t) = i(t)R_3 = 6e^{-500t} \text{ volts}}$$

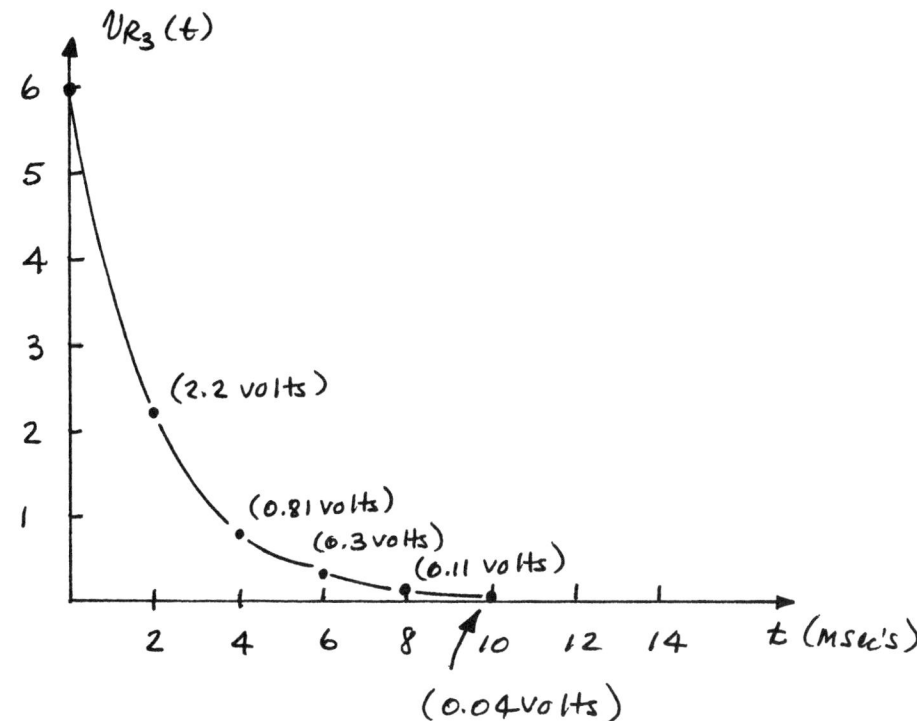

5-16

In the following circuit, the switch closes at $t = 0$. Find $v_o(t)$ for $t \geq 0$.

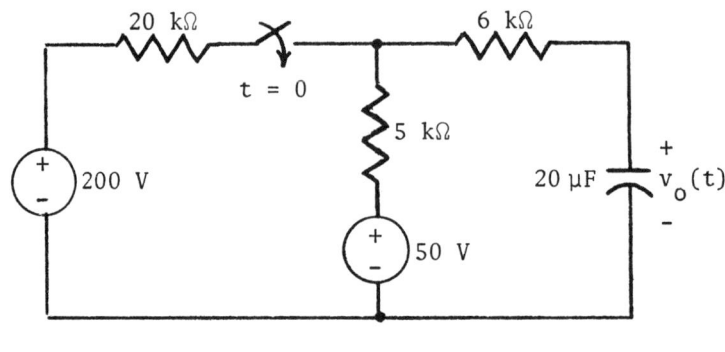

***

For this single time-constant circuit,

$$V_o(t) = V(\infty) + [V(0+) - V(\infty)] e^{-t/\tau} \text{ V}, \quad t \geq 0$$

in which $V(0+)$ is the initial capacitor voltage, $V(\infty)$ is the final capacitor voltage, and τ is the circuit time-constant $R_{TH}C$.

Since a capacitor voltage cannot jump, the initial capacitor voltage is equal to the capacitor voltage immediately before the switch is closed. By inspection, this is 50V: $V(0+) = 50$ V. A long time after the switch is closed, the capacitor will appear to be an open circuit to the dc excitation. Then, there will be no current through the 6-kΩ resistor and so no voltage across it. Consequently, $V(\infty)$ is equal to the voltage drop, top to bottom, across the series combination of the 5-kΩ resistor and the 50-V source. By nodal analysis,

$$\frac{V(\infty) - 50}{5000} + \frac{V(\infty) - 200}{20000} = 0$$

which solves to $V(\infty) = 80$ V. Also, R_{Th} is the resistance "seen" by the capacitor with the sources "killed." It is

$$R_{Th} = 6 + 5 \| 20 = 6 + \frac{5(20)}{5+20} = 6+4 = 10 \text{ k}\Omega$$

And so

$$\tau = R_{Th}C = (10^4)(20 \times 10^{-6}) = 0.2 \text{ s}$$

Finally, $V_0(t) = V(\infty) + [V(0+) - V(\infty)] e^{-t/\tau}$
$= 80 + [50 - 80] e^{-t/0.2}$
$= 80 - 30 e^{-5t}$ V, $t \geq 0$

5-17

For the following R-L circuit, find i as a function of time and sketch i. First find the initial condition, then develop a descriptive differential equation and solve.

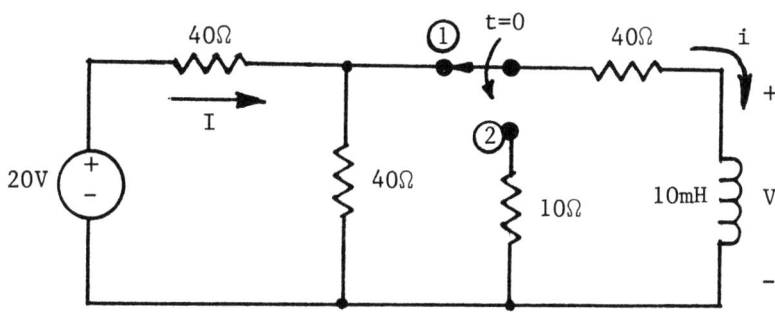

****

Since The Inductor Is a Short Circuit For $t < 0$, The Initial Current, I, IS

$$I(0^-) = 20/(40 + 40//40) = 20/60 = 1/3 \text{ A}.$$

If Current Division Is Used,

$$i(0^+) = i(0^-) = 1/3 \cdot 40/(40+40) = 1/6 \text{ A}.$$

After The Switch Moves From Point ① to Point ③, A Simple Mesh is Formed.

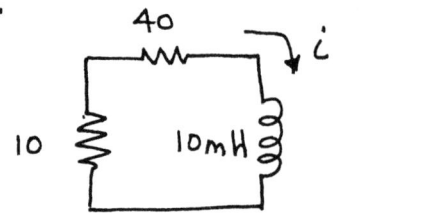

$i(0+) = 1/6 \, A$

From Kirchhoff's Voltage Law,
$$10i + 40i + 10 \times 10^{-3} \, di/dt = 0$$
or
$$di/dt + 5000 \, i = 0$$

IF $i = Ae^{st}$ Is Assumed As A Solution, then The Last Equation Becomes
$$A s e^{st} + 5000 \, A e^{st} = 0$$
or
$$s = -5000$$

Thus, $\quad i(t) = Ae^{st}$

And $\quad i(0+) = 1/6 = Ae^0$

This Gives $\quad A = 1/6.$

Now $\quad i(t) = \frac{1}{6} e^{-5000t} \, A.$

The Sketch of $i(t)$ Becomes

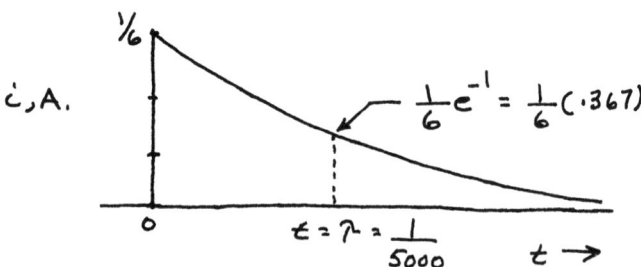

5-18

The switch has been closed for a long time and opens at t = 0. Determine $i(t)$, $t \geq 0$.

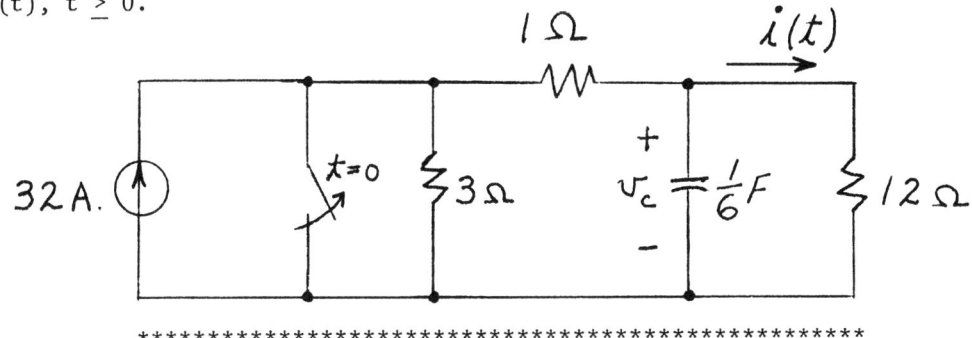

$i(0^+) = i(0^-) = 0$ SINCE $v_c(0^+) = v_c(0^-) = 0$

THEREFORE, SINCE THE ONLY SOURCE IS A CONSTANT, $i(t)$ MUST BE OF THE FORM

$$i(t) = A + Be^{-t/\tau}$$

WHERE $\tau = R_{eq} C = [4 \| 12](\frac{1}{6}) = \frac{1}{2}$ SEC.

HENCE, $i(t) = A + Be^{-2t}$

NOW, $i(0^+) = 0 = A + B$

AND $\lim_{t \to \infty} i(t) = A = \frac{3}{3+1+12}(32) = 6$

SOLVING GIVES $A = 6$ AND $B = -6$

AND $i(t) = 6(1 - e^{-2t})$ A., $t \geq 0$

5-19

For the circuit shown below find and sketch $i_L(t)$.

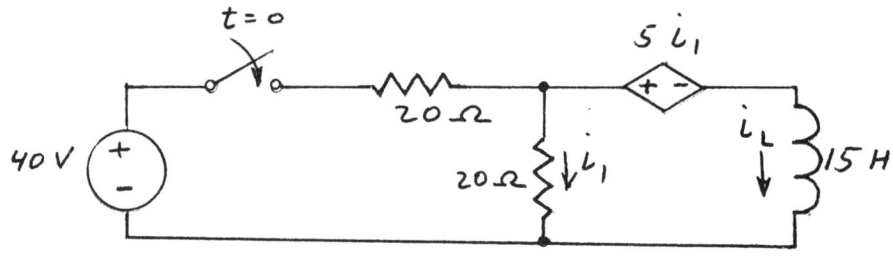

***

CAN REPLACE CIRCUIT TO LEFT OF INDUCTOR
WITH IT'S THEVENIN EQUIVALENT FOR $t \geq 0$

$$V_{th} = \frac{40 \cdot 20}{20+20} - 5 \cdot \frac{40}{20+20} = 15V \text{ (OPEN CIRCUIT VOLTAGE)}$$

TO FIND R_{th} "KILL" 40 V SOURCE AND
REPLACE INDUCTOR WITH 1 AMP SOURCE.
THEN

$$R_{th} = V_0 = 1 \cdot \frac{20 \cdot 20}{20+20} - 5 \cdot \frac{20}{20+20} = 7.5 \Omega$$

SO CIRCUIT BECOMES:

FORCED $i_L = i_{Lf} = \frac{15}{7.5} = 2A$

NATURAL $i_L = i_{Ln} = Ae^{-t/L/R} = Ae^{-\frac{t}{2}}$

$i_L = Ae^{-t/2} + 2$

$i_L(0) = 0 = A + 2 \quad ; \quad A = -2$

$\therefore i_L = 2(1 - e^{-t/2})$ AMPS $t \geq 0$ ANS.

$= 0 \quad t < 0$

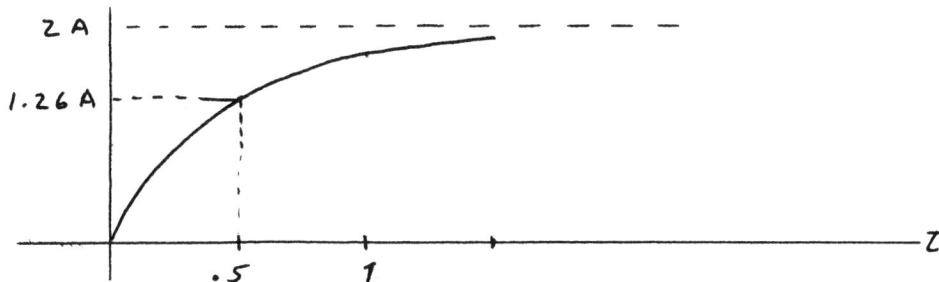

5-20

After switch closure the time constant is one second. What is the value of R?

****

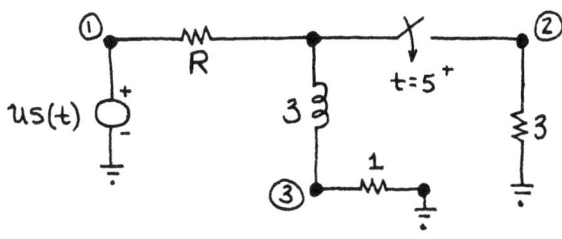

$R_{TH}(2,3) = 3||R + 1 = \dfrac{3R}{3+R} + 1 = \dfrac{3+4R}{3+R}$

$L/R_{TH}(2,3) = 1$; then $3 = R_{TH}(2,3) = \dfrac{3+4R}{3+R}$

which yields $R = 6$ ohms.

5-21

Find $v_c(t)$.

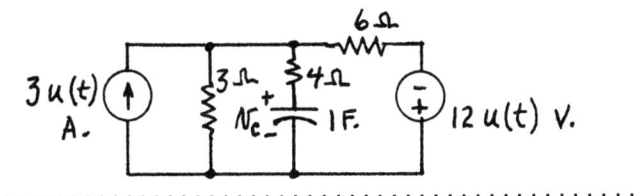

WE CAN WRITE KCL, OR THEVENIZE ABOUT C:

REDUCING, (for $t>0$)

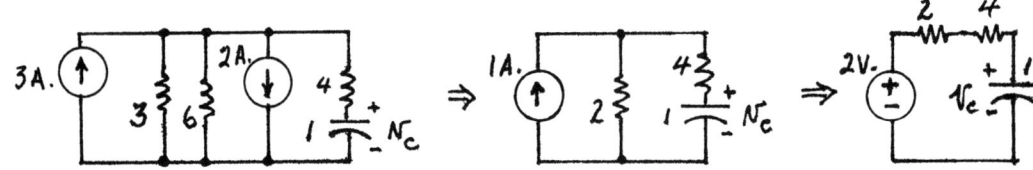

BY KCL:

(1) $\dfrac{dv_c}{dt} + \dfrac{v_c - 2}{6} = 0$

$\dfrac{dv_c}{dt} + \dfrac{1}{6} v_c = \dfrac{1}{3}$

THEN, $v_c = Ke^{-\frac{t}{6}} + A$, HERE, $A = 2$, $v_c(0^+) = 0$

$v_c = Ke^{-\frac{t}{6}} + 2$

$v_c(0^+) = 0 = K + 2$, $K = -2$

$\therefore v_c(t) = 2\left(1 - e^{-\frac{t}{6}}\right)$

5-22

For the given circuit, determine the capacitor initial condition voltage, the capacitor steady state voltage, the circuit time constant, and the capacitor time domain voltage expression.

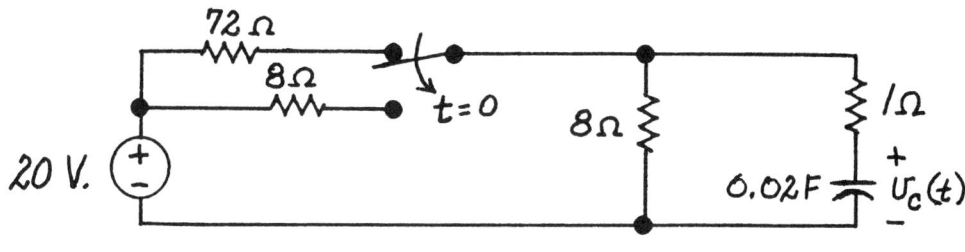

***

$U_c(0) = \left(\frac{8}{80}\right) 20 = \underline{2V}.$ CAPACITOR VOLTAGE SAME AS VOLTAGE ACROSS 8Ω RESISTOR BEFORE SWITCH IS THROWN.

$U_c(+\infty) = \left(\frac{8}{8+8}\right) 20 = \underline{10V}.$ CAPACITOR STEADY-STATE VOLTAGE AFTER SWITCH IS THROWN.

$T = R_{eq} C = (5)(0.02)$ $R_{eq} = \frac{(8)(8)}{8+8} + 1 = 5\Omega$

$\underline{T = \frac{1}{10} sec.}$ $U_{ss} = U_c(+\infty) \ \& \ U_{tr} = k e^{-t/\tau}$

$U_c(t) = U_{tr} + U_{ss} = 10 + k e^{-10t}; \ t \geq 0$

$U_c(0) = 2 = 10 + k e^{0} \Rightarrow k = -8$

$\underline{U_c(t) = 10 - 8 e^{-10t} \ V.; \ t \geq 0}$

5-23

Find the voltage, v, as shown for all time t > 0 if $v_1(0) = 1$ V and $v_2(0) = 2$V as shown.

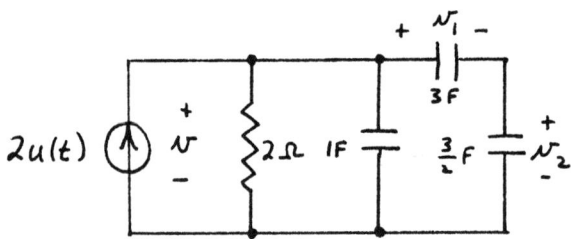

****

Reduce the capacitive network to a single capacitor. The series capacitors of $3F$ and $\frac{3}{2}F$ becomes

$$C_s = \frac{3 \times 3/2}{3 + 3/2} = \frac{9/2}{9/2} = 1F$$

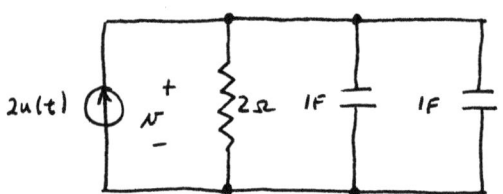

The parallel combination of two $1F$ capacitors becomes $2F$:

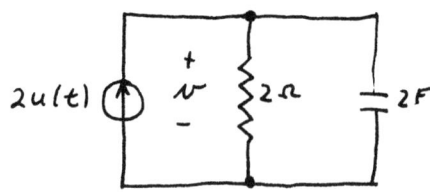

Kirchoff's current law yields:

$$2\frac{dv}{dt} + \frac{v}{2} = 2 \qquad \text{for } t > 0.$$

$$\frac{dv}{dt} + \frac{v}{4} = 1$$

The particular solution must be a constant, $v_p(t) = K$. Substitute this into the equation above to get:

$$0 + \frac{K}{4} = 1, \quad K = 4$$

The complementary solution is of the form:

$$v_c(t) = Ce^{-t/4} = Ce^{-t/RC}$$

$$v(t) = v_c(t) + v_p(t)$$

$$v(t) = Ce^{-t/4} + 4$$

From the original circuit, $v(0) = v_1(0) + v_2(0)$

$$v(0) = 3V = C + 4$$

$$C = -1$$

$$v(t) = 4 - e^{-t/4}$$

5-24

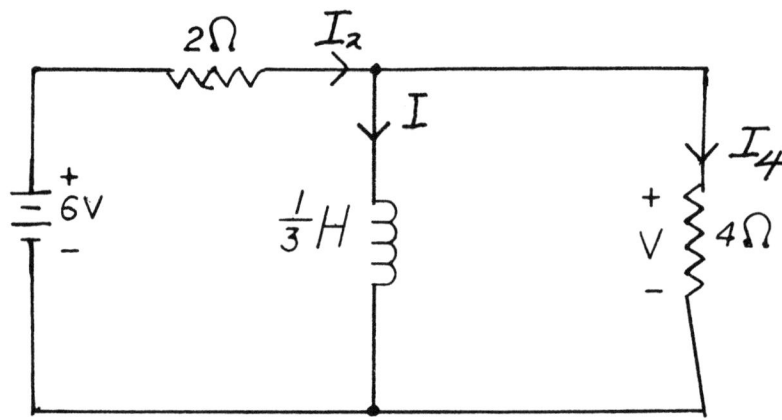

In this circuit v(0-) = 2 volts. Determine I(t) for t greater than 0.

****

First obtain the value of I at $t=0-$.
At $t=(0-)$, $V = 2$ Volts and $I_4 = 2/4 = 0.5$ Amps.
$I_2 = \frac{6-2}{2} = 2$ Amps at $t = 0-$
Using the current law at the top right hand node, $I = I_2 - I_4 = 2 - 0.5 = 1.5$ Amps.

Obtain the differential equation for I.
Voltage law around left mesh,
$2 I_2 + \frac{1}{3} \frac{dI}{dt} = 6$ and around the outer loop. $2 I_2 + 4 I_4 = 6$, $I_4 = I_2 - I$
$2 I_2 + 4 I_2 - 4 I = 6$, $6 I_2 = 4I + 6$
and $I_2 = 2/3 I + 1$. Substitute this in the left mesh equation.

$$\frac{4}{3}I + 2 + \frac{1}{3}\frac{dI}{dt} = 6 \quad \text{Multiply by 3.}$$

$$\frac{dI}{dt} + 4I = 12$$

The solution is:

$$I = Ae^{-pt} + B = \underline{Ae^{-4t} + 3 \text{ Amps}}$$

Alternate Solution. $I = I_N + I_F$

For natural response the two resistors are in parallel and

$$R_{eq.} = \frac{2 \times 4}{2+4} = \frac{4}{3} \quad \text{and} \quad \tau = \frac{L}{R_{eq.}} = \frac{1}{4}$$

For forced response the inductance is a short circuit after a long time and $I_F = 6/2 = 3$ Amps, $I_N = Ae^{-t/\tau}$

$$\underline{I = Ae^{-4t} + 3 \text{ Amps.}} \quad \text{as above.}$$

To determine A, use initial value of I which is 1.5 Amps at $t = 0^-$ and at $t = 0^+$ by continuity of induction current.

$$1.5 = A + 3, \quad A = -1.5$$

$$\underline{I = -1.5 e^{-4t} + 3 \text{ Amps.}}$$

5-25

Find the current, i(t), flowing in the inductor as shown for all time, t ≥ 0. The circuit is originally unenergized and v(t) = 25[sint . u(t) - Sint . u(t - π)].

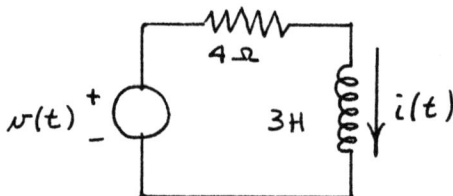

***

For $0 \le t \le \pi$, $3\frac{di}{dt} + 4i = 25 \sin t$.

The particular solution is:

$$i_p(t) = K_1 \sin t + K_2 \cos t$$

Substitute into the differential equation to get:

$$3K_1 \cos t - 3K_2 \sin t + 4K_1 \sin t + 4K_2 \cos t = 25 \sin t$$

$3K_1 + 4K_2 = 0$ $\qquad K_2 = -\frac{3}{4} K_1$

$4K_1 - 3K_2 = 25$ $\qquad 4K_1 + \frac{9}{4} K_1 = \frac{25}{4} K_1 = 25$

$K_1 = 4$ $\qquad K_2 = -3$

$$i_p(t) = 4 \sin t - 3 \cos t$$

The complementary solution is:

$$i_c(t) = C_1 e^{-4t/3} = C_1 e^{-\frac{R}{L}t}$$

$$i(t) = C_1 e^{-4t/3} + 4 \sin t - 3 \cos t$$

$$i(0) = 0 = C_1 - 3, \qquad C_1 = 3$$

$$i(t) = 3e^{-4t/3} + 4\sin t - 3\cos t$$

For $t > \pi$, $\quad 3\dfrac{di}{dt} + 4i = 0$

$$i(t) = C_2 e^{-4t/3}$$

$$i(\pi) = 3e^{-4\pi/3} + 4\sin\pi - 3\cos\pi = C_2 e^{-4\pi/3}$$

$$C_2 = 3 + 3e^{4\pi/3}$$

$$i(t) = (3 + 3e^{4\pi/3})e^{-4t/3} = 201 e^{-4t/3}$$

$$= 3e^{-4t/3} + 3e^{-4(t-\pi)/3}$$

SECOND ORDER CIRCUITS

5-26

The following differential equation represents a mesh current in a RLC circuit. Determine the characteristic equation, the natural frequencies, and the "form" of the natural and forced responses.

$$3\dfrac{d^2i}{dt^2} + 12\dfrac{di}{dt} + 24i = t + 3e^{-3t} \text{ A.; } t \geq 0$$

****

REWRITE THE D.E.: $\dfrac{d^2i}{dt^2} + 4\dfrac{di}{dt} + 8i = \tfrac{1}{3}t + e^{-3t}$ A.; $t \geq 0$

CHARACTERISTIC EQN.: $\underline{s^2 + 4s + 8 = 0}$

NATURAL FREQUENCIES: $s_{1,2} = \dfrac{-4 \pm \sqrt{16 - (4)(8)}}{2} = \underline{-2 \pm j2}$

NATURAL RESPONSE "FORM": $i_n(t) = k_1 e^{s_1 t} + k_2 e^{s_2 t}$

$\Rightarrow \underline{i_n(t) = e^{-2t}(k_3 \cos 2t + k_4 \sin 2t)}$

FORCED RESPONSE "FORM": $\underline{i_f(t) = k_5 + k_6 t + k_7 e^{-3t}}$

5-27

The capacitor voltage is 12 volts before the switch is closed in the circut below, the switch is closed at t = 0. Find the capacitor voltage, v(t), for all time $t \geq 0$.

***

$$C\frac{dv}{dt} + \frac{1}{L}\int v\, dt = 0$$

$$\frac{1}{4}\frac{dv}{dt} + 6.25 \int v\, dt = 0$$

$$\frac{dv}{dt} + 25 \int v\, dt = 0$$

$$\frac{d^2v}{dt^2} + 25v = 0$$

$$v(t) = C_1 \sin 5t + C_2 \cos 5t$$

$$v(0) = 12 = C_2$$

$$v(t) = C_1 \sin 5t + 12 \cos 5t$$

Also from the circuit

$$C\frac{dv}{dt} = i(t) \qquad \text{so,}$$

$$C\frac{dv}{dt}\bigg|_{t=0} = i(0) = 0, \quad \frac{dv}{dt}\bigg|_{t=0} = 0$$

$$\frac{dv}{dt} = 5C_1 \cos 5t - 60 \sin 5t$$

$$\frac{dv}{dt}\bigg|_{t=0} = 5C_1 = 0, \quad C_1 = 0$$

$$v(t) = 12 \cos 5t$$

5-28

The voltage source in the network of the figure below is described by the equation $v_1 = K_1 t$ for $t \geq 0$ and $v_1 = 0$ for $t < 0$. Determine $v_2(t)$ by using differential equation.

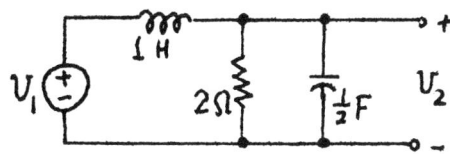

****

$$\frac{V_2}{2} + \frac{1}{2}\frac{dV_2}{dt} + \frac{1}{1}\int(V_2 - V_1)dt = 0$$

$$\frac{d^2V_2}{dt^2} + \frac{dV_2}{dt} + 2(V_2 - V_1) = 0 \; ; \; \frac{d^2V_2}{dt^2} + \frac{dV_2}{dt} + 2V_2 = 2K_1 t$$

$$s^2 + s + 2 = 0, \quad s = \frac{1}{2}[-1 \pm \sqrt{1-8}]$$

$$V_2(0^-) = V_2(0^+) = 0, \quad \frac{dV_2}{dt}(0^+) = 0, \quad \frac{d^2V_2}{dt^2}(0^+) = 0$$

V_{2c} = complementary solution = $e^{-0.5t}[A_1 \cos\frac{\sqrt{7}}{2}t + A_2 \sin\frac{\sqrt{7}}{2}t]$

V_{2p} = particular integral = $B_0 t + B_1$

$V_2(t)$ = complete solution = $V_{2c} + V_{2p}$

$$\frac{dV_2}{dt} = -0.5 e^{-0.5t}[A_1 \cos\frac{\sqrt{7}}{2}t + A_2 \sin\frac{\sqrt{7}}{2}t]$$
$$+ e^{-0.5t}[-\frac{A_1\sqrt{7}}{2}\sin\frac{\sqrt{7}}{2}t + \frac{\sqrt{7}}{2}A_2 \cos\frac{\sqrt{7}}{2}t] + B_0$$

At $t = 0$, $0 = A_1 + B_1 \; ; \quad 0 = -0.5 A_1 + \frac{\sqrt{7}}{2}A_2 + B_0$

At $t \to \infty$, the transient term (complementary solution) disappears $B_0 + 2B_0 t + 2B_1 = 2K_1 t$

from which $B_0 + 2B_1 = 0$, $2B_0 = 2K_1$. Thus $B_0 = K_1, B_1 = -\frac{K_1}{2}$

Also $A_1 = -B_1 = \frac{K_1}{2}$, $A_2 = -\frac{3}{2\sqrt{7}}$

5-29

The circuit below is initially unenergized, find i(t) for all time $t \geq 0$.

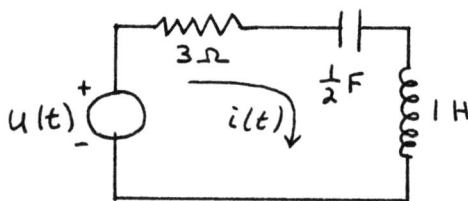

***

$$3i + 2\int i\, dt + \frac{di}{dt} = 1 \quad , \text{ for } t \geq 0$$

$$\frac{d^2i}{dt^2} + 3\frac{di}{dt} + 2i = 0$$

Using operator formalism:

$$s^2 + 3s + 2 = 0$$

$$s = \frac{-3 \pm \sqrt{9-8}}{2} = \frac{-3 \pm 1}{2} = -2, -1$$

$$\therefore i(t) = C_1 e^{-t} + C_2 e^{-2t}$$

The initial conditions are: $i(0) = 0$ and

$$3i(0) + 2v_c(0) + \frac{di}{dt}\bigg|_{t=0} = 1$$

but $v_c(0) = 0$ and $i(0) = 0$, so

$$\frac{di}{dt}\bigg|_{t=0} = 1$$

$$\frac{di}{dt} = -C_1 e^{-t} - 2C_2 e^{-2t}$$

$$\frac{di}{dt}\bigg|_{t=0} = -C_1 - 2C_2 = 1$$

$$i(0) = C_1 + C_2 = 0 \qquad C_2 = -C_1, \; C_1 = 1$$

$$C_2 = -1 \qquad \therefore i(t) = e^{-t} - e^{-2t}$$

5-30

Find the initial value for $V_C(0^+)$ and $V_L(0^+)$ for the following circuit.

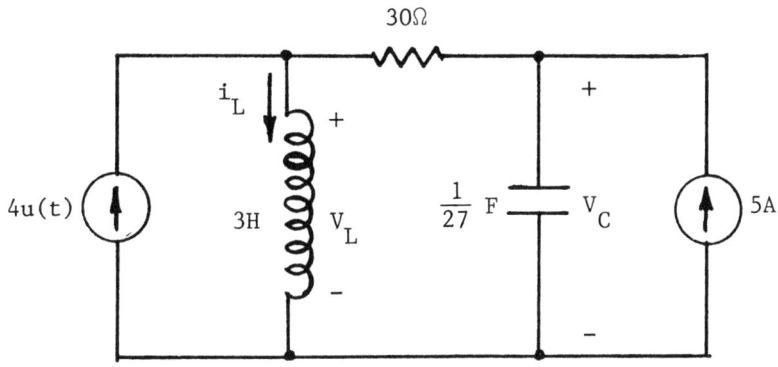

****

For $t = 0^-$, the Circuit Is As Follows:

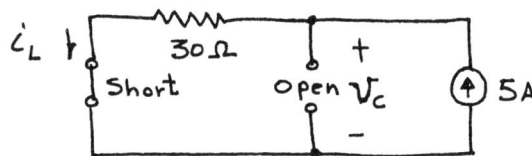

Thus, the Capacitor Voltage Is

$$V_C(0^+) \equiv V_C(0^-) = 5 \cdot 30 = 150 \text{ V}.$$

For $t = 0^+$, $V_L(0^+) \neq V_L(0^-)$. Hence, The Circuit For $t = 0^+$ Becomes

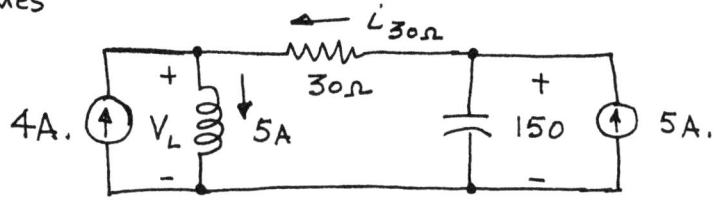

By Kirchhoff's Current Law

$$i_{30\Omega} + 4 - 5 = 0 \quad \text{or} \quad i_{30\Omega} = 1 \text{A}.$$

Then By Kirchhoff's Voltage Law

$$-150 + i_{30\Omega} \cdot 30 + V_L = 0$$

Or

$$V_L = 150 - 30 \cdot 1 = 120 \text{ V}.$$

5-31

Find the voltage, v, as shown. The circuit is initially unenergized.

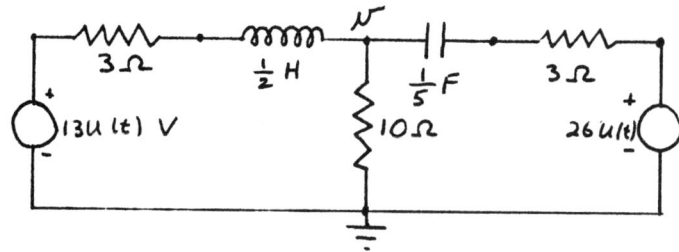

****

For $t \geq 0$

$$13 i_1 + \tfrac{1}{2} \frac{di_1}{dt} - 10 i_2 = 13$$

$$-10 i_1 + 13 i_2 + 5 \int i_2 dt = -26$$

$$10 i_1 = 13 i_2 + 5 \int i_2 dt + 26$$

$$i_1 = 1.3 i_2 + 0.5 \int i_2 dt + 2.6$$

$$13 i_1 = 16.9 i_2 + 6.5 \int i_2 dt + 33.8$$

$$\frac{di_1}{dt} = 1.3 \frac{di_2}{dt} + 0.5 i_2$$

$$\tfrac{1}{2} \frac{di_1}{dt} = 0.65 \frac{di_2}{dt} + 0.25 i_2$$

$$16.9 i_2 + 6.5 \int i_2 dt + 33.8 + 0.65 \frac{di_2}{dt} + 0.25 i_2 - 10 i_2 = 13$$

$$0.65 \frac{di_2}{dt} + 7.15 i_2 + 6.5 \int i_2 dt = -20.8$$

$$\frac{di_2}{dt} + 11 i_2 + 10 \int i_2 dt = 32$$

$$\frac{d^2 i_2}{dt^2} + 11\frac{di_2}{dt} + 10\, i_2 = 0$$

$$S^2 + 11S + 10 = (S+10)(S+1) = 0$$

$$i_2(t) = A e^{-10t} + B e^{-t}$$

$$i_1(t) = 1.3 A e^{-10t} + 1.3 B e^{-t} + .5A \int e^{-10t} dt + .5B \int e^{-t} dt + 2.6$$

$$= 1.25 A e^{-10t} + 0.8 B e^{-t} + C + 2.6$$

$$\lim_{t \to \infty} i_1(t) = 1 \qquad \therefore \quad C = -1.6$$

$$i_1(t) = 1.25 A e^{-10t} + 0.8 B e^{-t} + 1$$

$$i_1(0) = 0 = 1.25 A + 0.8 B + 1$$

$$i_2(0) = -2 = A + B$$

$$A + B = -2 \qquad 1.25 A + 0.8 B = -1$$

$$\begin{aligned} 1.25 A + 1.25 B &= -2.5 \\ -1.25 A - 0.8 B &= +1.0 \end{aligned} \qquad \therefore B = -\frac{10}{3}$$

$$A = \frac{4}{3}$$

$$i_1(t) = \frac{5}{3} e^{-10t} - \frac{8}{3} e^{-t} + 1$$

$$i_2(t) = \frac{4}{3} e^{-10t} - \frac{10}{3} e^{-t}$$

$$v(t) = 10(i_1 - i_2) =$$

$$10\left(\frac{5}{3} e^{-10t} - \frac{8}{3} e^{-t} + 1 - \frac{4}{3} e^{-10t} + \frac{10}{3} e^{-t}\right)$$

$$= 10\left(\frac{1}{3} e^{-10t} + \frac{2}{3} e^{-t} + 1\right)$$

$$= \frac{10}{3} e^{-10t} + \frac{20}{3} e^{-t} + 10$$

5-32

Given the circuit shown, which has the unit step response shown

A. Write the transfer function relating V_2 to V_1

B. If L is one Henry and C is one microfarad, observe the step response and calculate R.

C. What value of R would lead to critical damping in the step response?

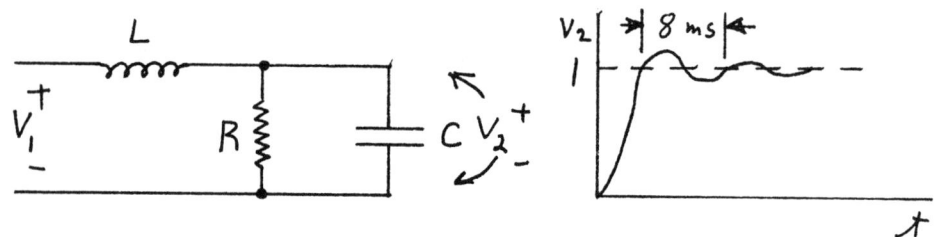

****

A.

S-domain impedances are sL, R and $\frac{1}{sC}$. The parallel impedance is

$$\frac{R\left(\frac{1}{sC}\right)}{R+\frac{1}{sC}} = \frac{R}{1+sCR}$$

Now we can apply voltage division to relate $V_2(S)$ to $V_1(S)$

$$\frac{V_2(S)}{V_1(S)} = \frac{\frac{R}{1+sCR}}{sL + \frac{R}{1+sCR}} = \frac{R}{s^2 LCR + sL + R}$$

B.

We see that the step response has damped sinusoids. Hence, the poles of the transfer function are complex.

$$S = \frac{-L \pm \sqrt{L^2 - 4CLR^2}}{2LCR} = \frac{-1}{2RC} \pm \sqrt{\frac{1}{LC} - \frac{1}{(2RC)^2}}$$

Since the period of oscillation is 8 ms,

$f = 1/8 \times 10^{-3} = 125$ Hz ; $\therefore \omega = 250\pi$

$$250\pi = \sqrt{\frac{1}{LC} - \frac{1}{(2RC)^2}} = \sqrt{\frac{1}{1\times 10^{-6}} - \frac{10^{12}}{4R^2}}$$

$$\left(\frac{10^6}{2R}\right)^2 = 10^6 - (250\pi)^2 = 383,149$$

$$R = \frac{5 \times 10^5}{\sqrt{383,149}} = \boxed{807.8}$$

C. Critical damping occurs when the quantity under the radical is equal to zero.

$$\frac{1}{(2RC)^2} = \frac{1}{LC} \; ; \; R^2 = \frac{LC}{4C^2} = \frac{1}{2}\sqrt{\frac{L}{C}} = \frac{1}{2}\sqrt{\frac{1}{10^{-6}}}$$

$$= \frac{1}{2}\sqrt{10^6} = \frac{1000}{2} = \boxed{500\,\Omega}$$

5-33

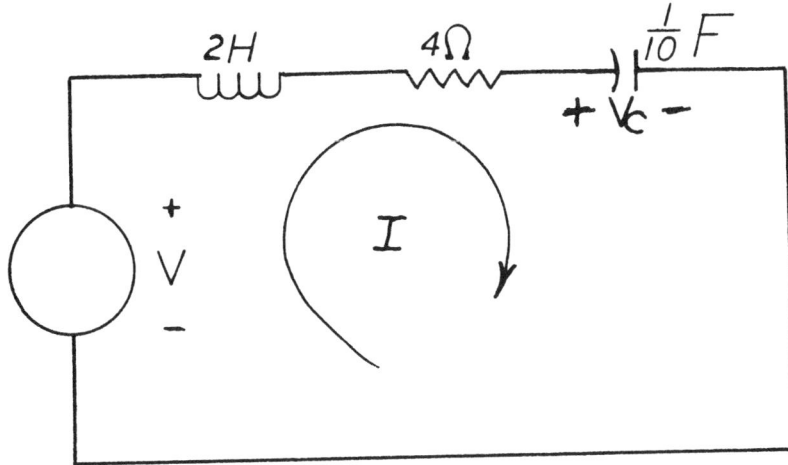

In this circuit $I(0) = 6A$, $\frac{dI}{dt}(0) = 10$ A/s, $V = 40 \cos t$. Find the complete response for $I(t)$.

****

Voltage Law equation around the circuit.

$$L\frac{dI}{dt} + RI + \frac{1}{C}\int_0^t I\,dt + V_c(0) = V$$

Differentiate and divide by L.

$$d^2I/dt^2 + \frac{R}{L}\frac{dI}{dt} + \frac{1}{LC}I = \frac{1}{L}\frac{dV}{dt}$$

The auxiliary equation is $s^2 + \frac{R}{L}s + \frac{1}{LC} = 0$

$$s^2 + 2s + 5 = 0, \; s = \frac{-2 \pm \sqrt{4-20}}{2} = -1 \pm j2$$

Natural response is:
$$I_N = e^{-t}(C\cos 2t + D\sin 2t)$$

To find the forced response try $I_F = A\cos t + B\sin t$ and substitute in the differential equation.

$$\frac{dI}{dt} = -A\sin t + B\cos t, \quad \frac{d^2I}{dt^2} = -A\cos t - B\sin t$$

In the equation: $-A\cos t - B\sin t + 2B\cos t - 2A\sin t + 5A\cos t + 5B\sin t = \frac{1}{2}\frac{dV}{dt} = -20\sin t$

$4A + 2B = 0$, $B = -2A$ and $(4B - 2A) = -20 = -10A$
$A = 2$ and $B = -4$. Then

$$I = I_N + I_F = e^{-t}(C\cos 2t + D\sin 2t) + 2\cos t - 4\sin t$$

$$\frac{dI}{dt} = -e^{-t}(C\cos 2t + D\sin 2t) + e^{-t}(-2C\sin 2t + 2D\cos 2t) - 2\sin t - 4\cos t$$

At $t = 0$ $6 = C + 2$, $C = 4$
$10 = -C + 2D - 4$, $2D = 10 + C + 4 = 18$
$D = 9$.

$$I = e^{-t}(4\cos 2t + 9\sin 2t) + 2\cos t - 4\sin t, \text{ Amps.}$$

6
SINUSOIDAL ANALYSIS

SINUSOIDS, PHASORS, AND COMPLEX ALGEBRA

━━━ 6-1

A voltage is given in phasor form as V = 4 + j3 V at a frequency of 100 Hz. Write this voltage in the time domain and find the instantaneous value at times t_1 = 2 ms and t_2 = 5 ms.

****

Polar form: $V = 5\angle 0.644 \text{ rad } V$; $\omega = 2\pi f = 628.32 \text{ rad/s}$

$v(t) = \text{Re}\left[5 e^{j(628.32 t + 0.644)} \text{ rad}\right] = 5 \cos(628.32 t + 0.644) \text{ rad}$

$\omega t_1 = (628.32)(.002) = 1.2566 \text{ rad}$

$v(.002) = 5 \cos(1.2566 + 0.644) = -1.619 \text{ V}$

$\omega t_2 = 3.1416 \text{ rad}$; $v(.005) = 5 \cos 3.7856 = -3.9985 \text{ V}$

6-2

For the following voltage and current expressions, indicate whether the element involved is a capacitor, inductor, or resistor <u>and</u> find the value of the C, L, or R.

$$v = 36 \sin(754t + 80°)$$
$$i = 4 \sin(754t + 170°)$$

****

i leads v by $90°$, ∴ element is <u>capacitor</u>

$$|Z_c| = \frac{V_p}{i_p} \qquad Z_c = jX_c = j\left(\frac{1}{\omega C}\right)$$

$$|Z_c| = \frac{36 V}{4 A}$$

$$\underline{|Z_c| = 9\, \Omega}$$

from expressions, $\omega = 754$

$$|Z_c| = \frac{1}{\omega C}$$

∴ $C = \frac{1}{\omega |Z_c|}$

$$C = \frac{1}{(754)(9)}$$

$$\boxed{C = 147.4\, \mu f}$$

6-3

Convert $i = A_1 \sin\omega t + A_2 \cos\omega t$ into the forms

1) $i = M \sin(\omega t + \theta_s)$
2) $i = M \cos(\omega t - \theta_c)$

using $\sin(\omega t \pm \theta_s) = \sin\omega t \cos\theta_s \pm \cos\omega t \sin\theta_s$

and $\cos(\omega t \pm \theta_c) = \cos\omega t \cos\theta_c \mp \sin\omega t \sin\theta_c$.

****

1) $A_1 \sin\omega t + A_2 \cos\omega t = \sqrt{A_1^2 + A_2^2} \left[\dfrac{A_1 \sin\omega t}{\sqrt{A_1^2 + A_2^2}} + \dfrac{A_2 \cos\omega t}{\sqrt{A_1^2 + A_2^2}} \right]$

$i = \sqrt{A_1^2 + A_2^2} \left[\sin\omega t \cos\theta_s + \cos\omega t \sin\theta_s \right]$

$i = M \sin(\omega t + \theta_s)$

WHERE $\theta_s = \tan^{-1} \dfrac{A_2}{A_1}$ AND $M = \sqrt{A_1^2 + A_2^2}$

2) $A_1 \sin\omega t + A_2 \cos\omega t = \sqrt{A_1^2 + A_2^2} \left[\dfrac{A_1 \sin\omega t}{\sqrt{A_1^2 + A_2^2}} + \dfrac{A_2 \cos\omega t}{\sqrt{A_1^2 + A_2^2}} \right]$

$i = \sqrt{A_1^2 + A_2^2} \left[\sin\omega t \cos\theta_c + \cos\omega t \cos\theta_c \right]$

$i = M \cos(\omega t - \theta_c)$ WHERE

$\theta_c = \tan^{-1} \dfrac{A_1}{A_2}$ AND $M = \sqrt{A_1^2 + A_2^2}$

6-4

For the circuit shown, find V_a.

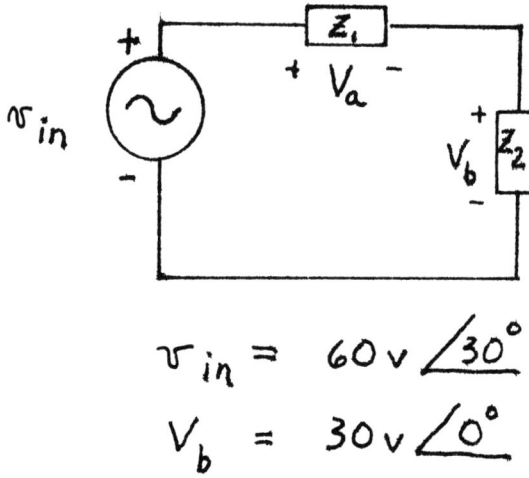

$$v_{in} = 60v \angle 30°$$
$$V_b = 30v \angle 0°$$

****

From Kirchoff's Voltage Law, $V_a = v_{in} - V_b$
Polar to rectangular coordinates yields:

$$v_{in} = 52 + 30j$$
$$V_b = 30 + 0j$$

$$V_a = v_{in} - V_b$$
$$V_a = (52 + 30j) - (30 + 0j)$$
$$V_a = 22 + 30j$$
$$V_a = 37.2v \angle 53.7°$$

6-5

In the following circuit which contains a sinusoidal 25-A rms current source, ammeter A_1 reads 15 A rms and ammeter A_2 reads 6 A rms. Find the reading of ammeter A_3.

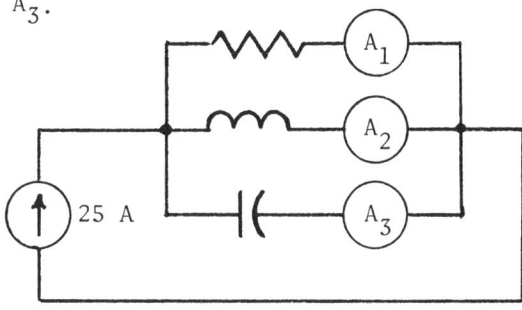

***

The reading of ammeter A_3 is **not**
$$25 - 15 - 6 = 4 \text{ A}$$
because scalar addition cannot be used. Instead, phasor addition must be used. For convenience we will reference the resistor current at $0°$. Then, because the inductor current lags the resistor current by $90°$ and the capacitor current leads it by $90°$, the phasor I_{LC} of the sum of the inductor and capacitor currents is the side opposite of one of the following two triangles:

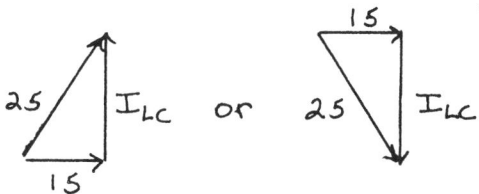

By Pythagoras's theorem,
$$I_{LC} = \sqrt{25^2 - 15^2} = 20 \text{ A}$$

Then, since the inductor and capacitor currents are $180°$ out of phase, either
$$|I_{LC}| = |I_L| - |I_c| \quad \text{or} \quad |I_{LC}| = |I_c| - |I_L|$$

and since $|I_L| = 6$ A, $|I_c|$ must be $20 + 6 = 26$ A to produce a positive $|I_{LC}|$. So, ammeter A_3 reads 26 A.

6-6

Given the circuit differential equation

$$\frac{d^4 i}{dt^4} + 4\frac{d^3 i}{dt^3} - 9\frac{d^2 i}{dt^2} - 35i = 100 \cos(2t - 40°)$$

use phasor concepts to determine the steady-state (sinusoidal) component of i(t).

****

We can transform the differential equation with the following transformations:

$$\frac{d^2}{dt^2} \to (j\omega)^2, \quad \frac{d^3}{dt^3} \to (j\omega)^3, \quad \frac{d^4}{dt^4} \to (j\omega)^4$$

$i \to I$ and $100 \cos(2t - 40) \to 100 \angle -40°$

with $\omega = 2$ rad/s. So, the differential equation transforms to

$$(j2)^4 I + 4(j2)^3 I - 9(j2)^2 I - 35 I = 100 \angle -40°$$

or
$$I = \frac{100 \angle -40°}{16 - j32 + 36 - 35} = \frac{100 \angle -40°}{17 - j32}$$

$$= \frac{100 \angle -40°}{36.2 \angle -62.0°} = 2.76 \angle 22.0° \text{ A}$$

And so $i(t) = 2.76 \cos(2t + 22.0°)$ A.

Note that this method requires much less effort than the conventional method of undetermined coefficients.

COMPLEX IMPEDANCE AND ADMITTANCE

6-7

Find V_2/V_1 for the circuit shown. (Values are impedances.)

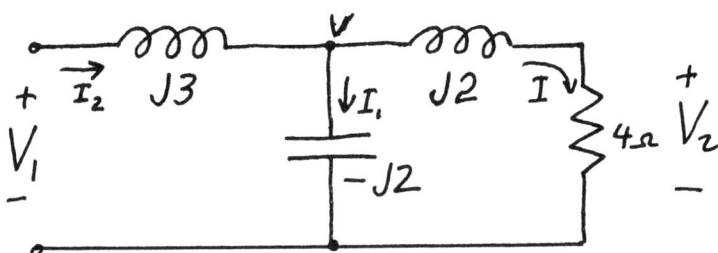

****

This problem is easily done using the "ladder" method. Assume $I = 1$ AMP then $V_2 = 4$ volts

$$V = 1(4+j2) = 4+j2 \text{ volts}$$

$$I_1 = \frac{4+j2}{-j2} = -1+j2 \text{ AMP}$$

$$I_2 = I_1 + I = -1+j2 + 1 = j2 \text{ AMP}$$

$$V_1 = V + I_2(j3) = 4+j2 + j2(j3) = -2+j2 \text{ volts}$$

$$V_2/V_1 = \frac{4}{-2+j2} \cdot \frac{-2-j2}{-2-j2} = \frac{-8-j8}{8} = -1-j$$

or $V_2/V_1 = \sqrt{2} \underline{/-135°}$

6-8

In the parallel RLC circuit, $\bar{V} = 100 \angle 0°$ V(RMS). Find L if $\bar{I} = (5 + j1)$A and $\omega = 10^4$ rad/sec.

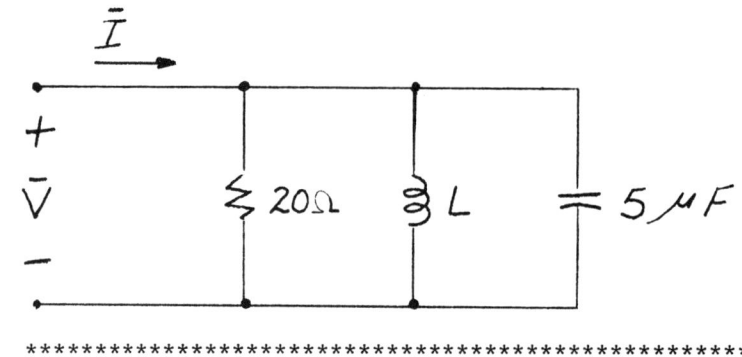

***

$$Y = \frac{\bar{I}}{\bar{V}} = \frac{5+j1}{100\angle 0°} = 0.05 + j\,0.01$$

THUS, $\quad \omega C - \dfrac{1}{\omega L} = 0.01$

AND $\quad L = \dfrac{1}{\omega(\omega C - 0.01)}$

SUBS. FOR $\omega = 10^4$ AND $C = 5 \times 10^{-6}$ GIVES

$$L = 2.5 \text{ mH}.$$

6-9

If v(t) = 100 cos (100t + 20°) and i(t) = 25 cos(100t - 10°),

a) Find Z, (b) Find the power supplied by the source,
c) What element (value and type) should be placed across the network so that the source voltage and current are in phase?

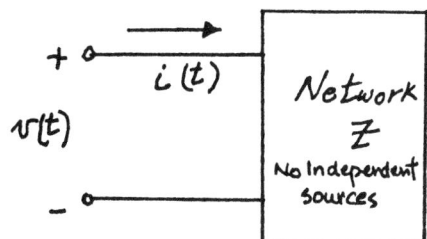

* * * * * * * * * * * * * * * * * * *

(a) Writing voltage and current as phasors and dividing

$$Z = \frac{100 \angle 20°}{25 \angle -10°} = 4 \angle 30°$$

(b) $P = V_{rms} I_{rms} \cos \phi = \frac{100}{\sqrt{2}} \cdot \frac{25}{\sqrt{2}} \cos(20+10)°$

$= 1082.5$ watts

(c) $Y = \frac{1}{Z} = .25 \angle -30° = .2165 - j.125$

To be pure real, add $Y = +j\, 0.125$ across netw.

$\therefore \frac{1}{\omega C} = 0.125$ Since $\omega = 100$, $C = .08$ farad

6-10

Determine the value of i(t) as shown in the steady state.

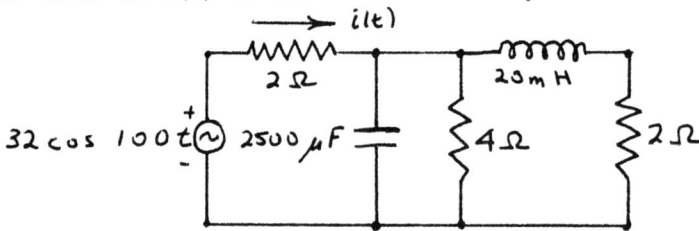

****

$$Y = \frac{j}{4} + \frac{1}{4} + \frac{1}{2(1+j)} = \frac{1}{4}(1+j) + \frac{1}{2}\left(\frac{1}{1+j}\right)$$

$$= \frac{1}{4}\left[\frac{(1+j)^2 + 2}{1+j}\right] = \frac{1}{4}\left[\frac{1+j2-1+2}{1+j}\right]$$

$$= \frac{1}{2}\left[\frac{1+j}{1+j}\right] = \frac{1}{2}$$

$$Z = \frac{1}{Y} = 2\,\Omega$$

$$I = \frac{32}{Z+2} = \frac{32}{4} = 8\text{ A}$$

$$i(t) = 8\cos 100t \quad A$$

6-11

Find the h-parameters for the circuit shown.

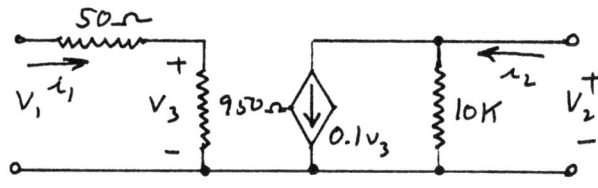

***

This is a somewhat simplified equivalent circuit for a bipolar junction transistor. h-parameter equations are:

$$V_1 = h_{11} i_1 + h_{12} V_2$$
$$i_2 = h_{21} i_1 + h_{22} V_2$$

We find h_{11} and h_{21} by making $V_2 = 0$, i.e. shorting the #2 terminals. Then

$$V_1 = i_1 (50 + 950) = 1000 i_1 \quad \therefore \boxed{h_{11} = 1000 \, \Omega}$$

Shorting the #2 terminals forces all the dependent source current through the short. $i_2 = 0.1 v_3$. But $v_3 = 950 i_1$, and so

$$i_2 = 0.1 \times 950 i_1 = 95 i_1 \quad \therefore \boxed{h_{21} = 95}$$

If we make $i_1 = 0$, things are very simple. The dependent current source is zero and no voltage is coupled into the #1 circuit. Hence $\boxed{h_{12} = 0}$

If V_2 is applied to #2 terminals, $i_2 = V_2/10K$

$$h_{22} = i_2/V_2 = 1/10K = \boxed{10^{-4}}$$

6-12

Find $\bar{Z}_{in}$ and give an equivalent series network.

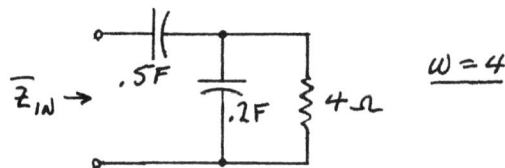

$\omega = 4$

***

$$\bar{Z}_{IN} = \frac{1}{j4(.5)} + \frac{4\left(\frac{1}{j4(.2)}\right)}{4 + \frac{1}{j4(.2)}} = \frac{1}{j2} + \frac{\frac{1}{.2j}}{4 + \frac{1.25}{j}}$$

$$= -.5j + \frac{5}{1.25 + j} = -.5j + \frac{6.25 - 5j}{2.56} = -.5j + 2.44 - 1.95j$$

$$\bar{Z}_{IN} = 2.44 - 2.45j \qquad \text{EQUIV. CIRC:}$$

2.44 Ω, .1F

6-13

Find the phasor current I and the phasor voltage drop across each element in polar form using the source as the reference for angles.

$f = 10,000\ Hz$, $10\angle 0$, 400 Ω, 0.03H, 0.01μF

***

$Z_R = 400\ \Omega$; $Z_L = j(20,000\pi)(.03) = j1885.0\ \Omega$

$Z_C = \frac{-j}{(20,000\pi)(0.01 \times 10^{-6})} = -j1591.5\ \Omega$

KVL: $-10 + 400I + j1885I - j1591.5I = 0$

$I = \frac{10\angle 0}{400 + j293.5} = 0.0202\angle -0.6330\ A$

$V_R = 400 I = 8.08 \angle{-0.6330}$ V; $V_L = j1885 I$
$V_L = 38.077 \angle{0.9378}$
$V_C = -j1591.5 I = 32.148 \angle{-2.204}$ V
$V_{source} = 10\angle 0$ V

6-14

Find the equation for current in the circuit.

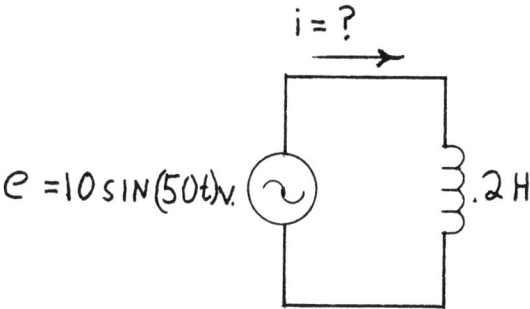

***

$\vec{E} = 10\angle 0°$ V.

$\vec{I} = \dfrac{\vec{E}}{\vec{Z_L}}$

$X_L = \omega L$
$= (50)(.2) = 10\ \Omega$
$\vec{Z_L} = 10j\ \Omega$

$\vec{I} = \dfrac{10\angle 0°}{10j} = 1\angle -90°$ AMPS.

$i = \underline{\underline{\sin(50t - 90°)}}$ VOLTS

PHASOR EQUIVALENT CIRCUITS

6-15

The circuit shown below can be used to make a resistor R appear to have the value aR ($0 \le a \le 1$). Find the values of X_L and X_C in terms of a and R.

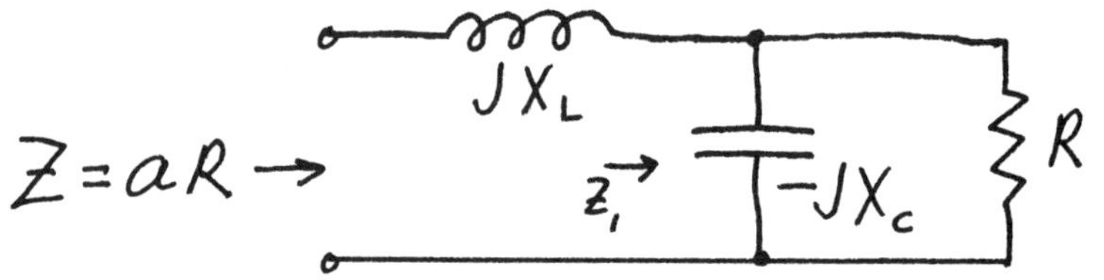

$Z = aR \rightarrow$ [circuit: jX_L in series, then parallel combination of $-jX_C$ and R; Z_1 indicated after jX_L]

****

$$Z_1 = \frac{R(-jX_C)}{R - jX_C} \cdot \frac{R + jX_C}{R + jX_C} = \frac{RX_C^2 - jR^2 X_C}{R^2 + X_C^2}$$

The real part of Z_1 must $= aR$

$$\frac{RX_C^2}{R^2 + X_C^2} = aR$$

$$X_C^2 = aR^2 + aX_C^2$$

$$X_C = R\sqrt{\frac{a}{1-a}}$$

The imaginary part of Z_1 must $= -X_L$

$$\frac{R^2 X_C}{R^2 + X_C^2} = X_L$$

using $X_C = R\sqrt{\frac{a}{1-a}}$

$$\frac{R^3 \sqrt{\frac{a}{1-a}}}{R^2 + \frac{aR^2}{1-a}} = X_L$$

$$X_L = \frac{R\sqrt{\frac{a}{1-a}}}{\frac{1}{1-a}}$$

$$X_L = R\sqrt{a(1-a)}$$

6-16

Prove that Kirchhoff's voltage law holds by drawing the appropriate phasor diagrams for the following circuit.

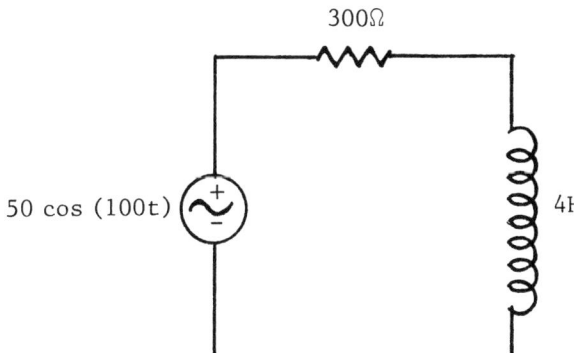

****

The Phasor Circuit Is As Follows:

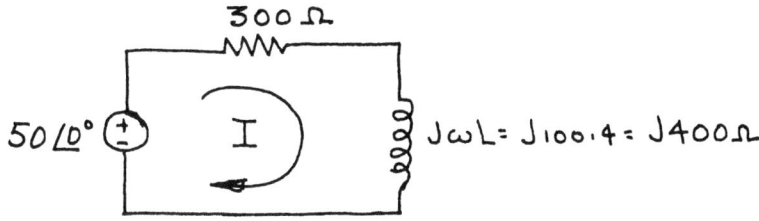

$$I = V/Z = 50/(300 + j400) = 50/[100(3+j4)]$$
$$= 0.5/(5\angle 53.1°) = 0.1\angle -53.1° \text{ A}.$$

$$V_{300} = IR = I\,300 = 300 \times 0.1\angle -53.1° = 30\angle -53.1° \text{ V}.$$

$$V_{4H} = I j\omega L = I j 400 = 0.1\angle -53.1° \times 400 \angle 90°$$
$$= 40 \angle (90 - 53.1°) = 40 \angle 36.9° \text{ V}.$$

The Vector Sum of V_{300} & V_{4H} Is

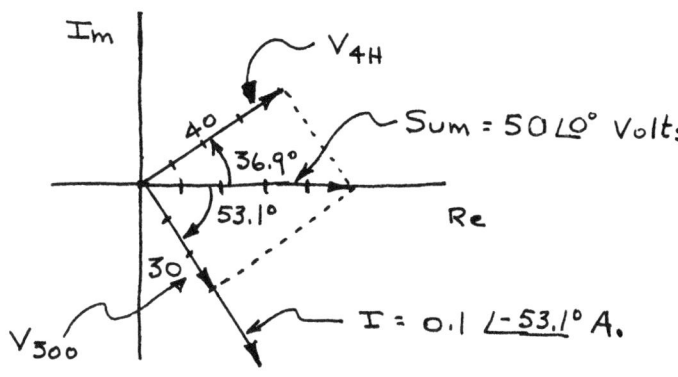

6-17

Determine the voltage v_L. The voltage source v_s has an rms value of 110V and a frequency of 159.15 cps.

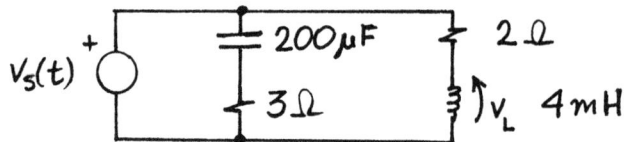

*****

$$X_L = 2\pi f \times L = 2\pi \times 159.15 \times 4 \times 10^{-3} = 4\,\Omega$$

$$Z = R + jX_L = 2 + j4 = 4.47\,\angle 63.4°$$

VOLTAGE DIVIDER RULE

$$V_L = V_s \frac{jX_L}{Z} = 110\angle 0° \times \frac{4\angle 90°}{4.47\angle 63.4°} = 98.4\angle 26.6°$$

NETWORK THEOREMS AND NODAL/MESH ANALYSIS

6-18

Determine the Thevenin equivalent impedance to the right of terminals a & b.

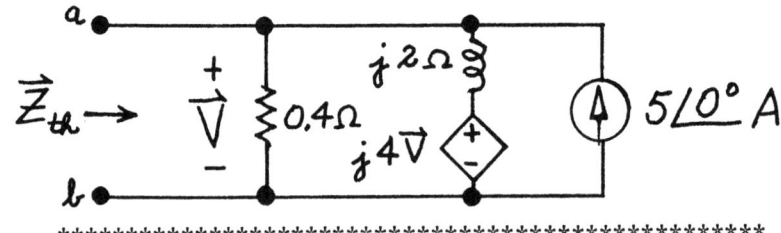

*****

$$\vec{I}_{sc} = \vec{I}_{ab} = 5\angle 0°\,A. \quad \& \quad \vec{V}_{o.c.} = \vec{V}_{ab} = \vec{V}$$

USE NODAL ANALYSIS.

$$\vec{Z}_{th} = \frac{\vec{V}_{o.c.}}{\vec{I}_{sc}} \qquad \frac{\vec{V}}{0.4} + \frac{\vec{V} - j4\vec{V}}{j2} = 5\angle 0° \Rightarrow \vec{V} = 7.07\angle 45°$$

$$\Rightarrow \vec{V}_{o.c.} = 7.07\angle 45° V$$

$$\Rightarrow \underline{\vec{Z}_{th} = 1.414\angle 45° = 1 + j1\,\Omega}$$

6-19

Use phasor analysis to determine the time domain sinusoidal steady state voltage at the terminals of the current source.

$i_g(t) = 3\cos(1250t - 15°)$ A.

****

CURRENT SOURCE PHASOR CURRENT: $\vec{I}_g = 3\angle -15°$ A.

CAPACITOR IMPEDANCE: $\vec{Z}_c = \frac{-j}{\omega C} = \frac{-j}{(1250)(10^{-4})} = -j8\,\Omega$

RESISTOR-CAPACITOR PARALLEL IMPEDANCE:

$$\vec{Z}_p = \frac{(6)(-j8)}{6 - j8} = 4.8\angle -36.87°\,\Omega$$

CURRENT SOURCE TERMINAL VOLTAGE:

$$\vec{V}_g = \vec{Z}_p \vec{I}_g = (4.8\angle -36.87°)(3\angle -15°) = 14.4\angle -51.87°\text{ V.}$$

$$\Rightarrow \underline{v_g(t) = 14.4\cos(1250t - 51.87°)\text{ V.}}$$

6-20

Write nodal equations for the circuit shown, and using determinants, find V_1.

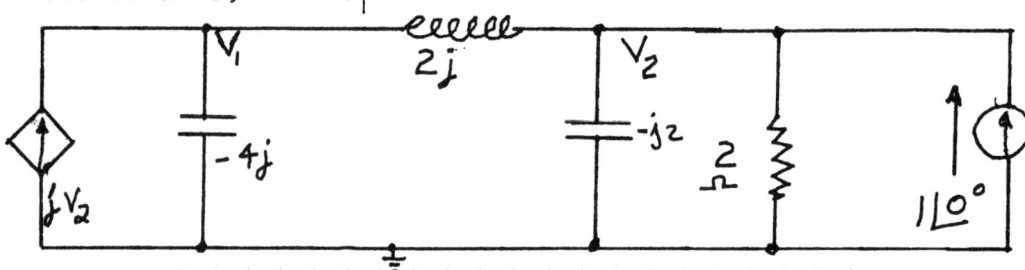

At the V_1 node: $\quad j V_2 = V_1 \left(\frac{1}{-4j} + \frac{1}{2j} \right) - V_2 \left(\frac{1}{2j} \right)$

$$0 = -\frac{jV_1}{4} - \frac{jV_2}{2}$$

$$0 = V_1 + 2V_2 \qquad \text{Eqn. (1)}$$

At the V_2 node: $\quad 1 = -V_1 \left(\frac{1}{2j} \right) + V_2 \left(\frac{1}{2j} + \frac{1}{-2j} + \frac{1}{2} \right)$

Simplifying, $\quad 2 = j V_1 + V_2 \qquad \text{Eqn. (2)}$

Determinant for Eqns (1) & (2)

$$\Delta = \begin{vmatrix} 1 & 2 \\ j & 1 \end{vmatrix} = 1 - 2j$$

$$V_1 = \frac{1}{\Delta} \begin{vmatrix} 0 & 2 \\ 2 & 1 \end{vmatrix} = -\frac{4}{1-2j} = -1.789 \, \underline{/63.43°}$$

$$= 1.789 \, \underline{/-116.57°}$$

6-21

Determine the phasor voltage V_B using nodal analysis. The current source is $8\angle 0°$ A and the voltage source is $10\angle 25°$ V.

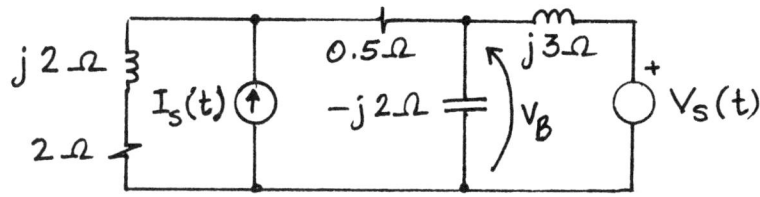

****

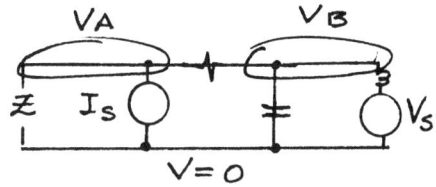

SELECTED NODES

$Z = 2 + j2 = 2.83\angle 45°$

KCL EQS AT NODES

A: $\quad \dfrac{V_A}{2.83\angle 45°} + \dfrac{V_A - V_B}{0.5} = 8\angle 0°$

B: $\quad \dfrac{V_B}{2\angle -90°} + \dfrac{V_B - V_A}{0.5} + \dfrac{V_B - V_S(t)}{3\angle 90°} = 0$

A: $\quad 2.26\angle -6.34°\, V_A - 2\angle 0°\, V_B = 8\angle 0°$

B: $\quad -2\angle 0°\, V_A + 2.01\angle 4.76°\, V_B = 3.33\angle -65°$

$$V_B = \dfrac{\begin{vmatrix} 2.26\angle -6.34° & 8\angle 0° \\ -2\angle 0° & 3.33\angle -65° \end{vmatrix}}{\begin{vmatrix} 2.26\angle -6.34° & -2\angle 0° \\ -2\angle 0° & 2.01\angle 4.76° \end{vmatrix}} = \dfrac{19.74\angle -21.2°}{0.56\angle -13.04°}$$

$V_B = 35.56\angle -8.2°$ V

6-22

Solve for the inductor current, $i(t)$, as shown using mesh equations.

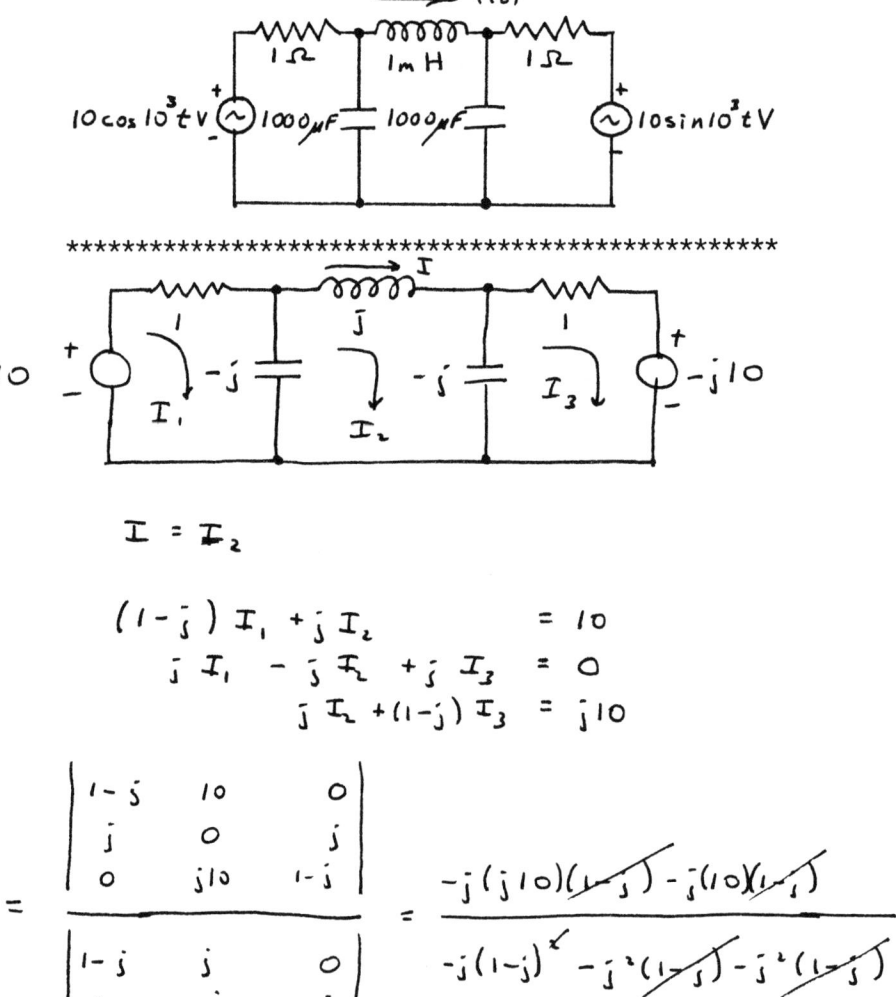

$$I = I_2$$

$$(1-j)I_1 + jI_2 = 10$$
$$jI_1 - jI_2 + jI_3 = 0$$
$$jI_2 + (1-j)I_3 = j10$$

$$I_2 = \frac{\begin{vmatrix} 1-j & 10 & 0 \\ j & 0 & j \\ 0 & j10 & 1-j \end{vmatrix}}{\begin{vmatrix} 1-j & j & 0 \\ j & -j & j \\ 0 & j & 1-j \end{vmatrix}} = \frac{-j(j10)(1-j) - j(10)(1-j)}{-j(1-j)^2 - j^2(1-j) - j^2(1-j)}$$

$$= \frac{10 - j10}{-j+1} = \frac{10(1-j)}{1-j} = 10$$

$$i(t) = 10 \cos 10^3 t$$

6-23

The frequency of each source is 1000 Hz with the phase relations shown on the diagram. Replace this circuit by its Norton phasor equivalent.

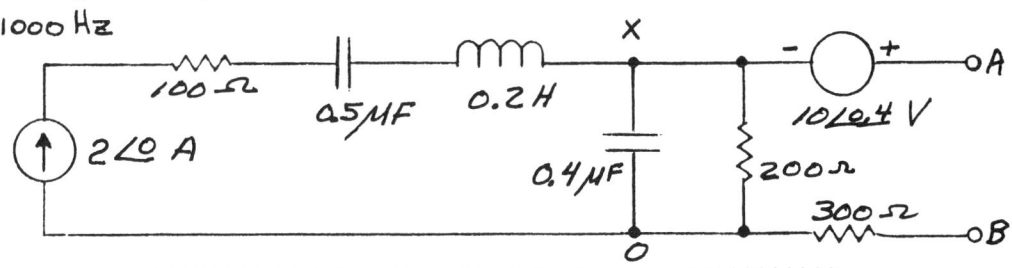

$f = 1000$ Hz

****

$Y_{0.4\mu F} = j(2000\pi)(0.4 \times 10^{-6}) = j0.0025133$ ℧

KCL Node X: $-2 + j0.0025133 \, V_{XO} + \frac{1}{200} V_{XO} + \frac{1}{300}\left(V_{XO} + 10\underline{/0.4}\right) = 0$

$V_{XO} = \dfrac{1.96934 \, \underline{/-0.0065913}}{0.008704 \, \underline{/0.29292}} = 226.257 \, \underline{/-0.29951}$

short circuit $I_{AB} = \dfrac{1}{300}\left[(216.184 - j66.7576) - 10\underline{/0.4}\right]$

$I_{AB} = 0.77997 \, \underline{/-0.27197}$ A

Replacing each source by its internal impedance:

$Z_{AB} = \dfrac{1}{0.005 + j0.0025133} + 300 = 459.66 - j80.254$

$Z_{AB} = 466.61 \, \underline{/-0.17285}$ Ω

$Y_{AB} = \dfrac{1}{Z_{AB}} = 0.0021431 \, \underline{/0.17285} = 0.0021093 + j0.0003686$

$R = \dfrac{1}{0.0021093} = 474.1$ Ω ; $C = \dfrac{0.0003686}{2000\pi} = 0.0586 \mu F$

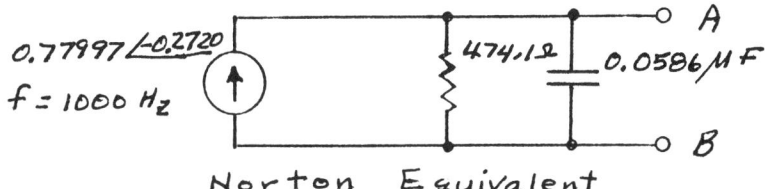

Norton Equivalent

6-24

Find the Thevenin's equivalent of the circuit to the left of a-b. The supply frequency is 60 Hz.

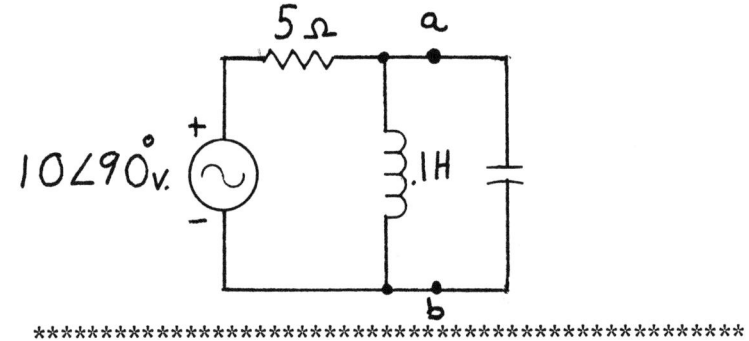

***

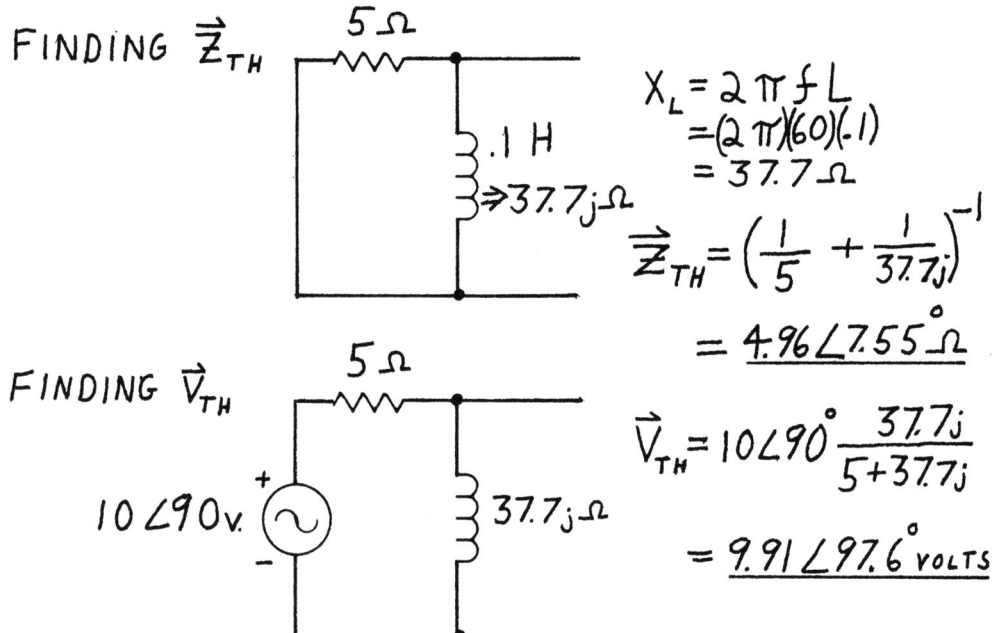

$$X_L = 2\pi f L = (2\pi)(60)(.1) = 37.7\,\Omega$$

$$\vec{Z}_{TH} = \left(\frac{1}{5} + \frac{1}{37.7j}\right)^{-1} = 4.96\angle 7.55°\,\Omega$$

$$\vec{V}_{TH} = 10\angle 90° \frac{37.7j}{5 + 37.7j} = 9.91\angle 97.6°\text{ VOLTS}$$

THE THEVENIN'S EQUIVALENT IS:

6-25

For the circuit shown, (a) construct the phasor domain equivalent circuit and (b) use the node voltage method to find the steady state voltage across the inductance.

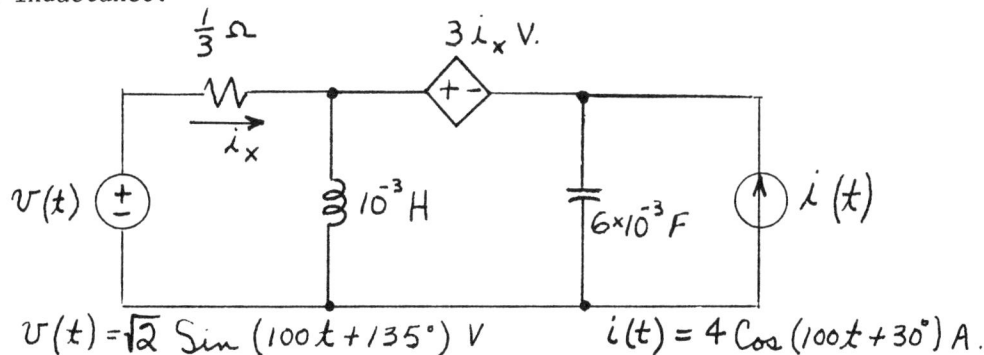

$v(t) = \sqrt{2} \sin(100t + 135°)$ V $\qquad i(t) = 4\cos(100t + 30°)$ A.

****

a)

FOR THE SUPERNODE, $\left(\underset{\text{Supernode}}{\Sigma} I_{out} = 0\right)$

$$\frac{\bar{V}_A - \sqrt{2}\angle 45°}{1/3} + \frac{\bar{V}_A}{j0.1} + \frac{\bar{V}_B}{-j\,5/3} - 4\angle 30° = 0$$

ALSO, $\bar{V}_A - \bar{V}_B = 3\bar{I}_x = 3\,\dfrac{\sqrt{2}\angle 45° - \bar{V}_A}{1/3} = 9(1+j1) - 9\bar{V}_A$

SOLVING GIVES $\bar{V}_A = 2.09\angle 137.29°$ V (RMS)

HENCE, $v_a(t) = 2.09\sqrt{2}\cos(100t + 137.29°)$ V.

AC STEADY STATE POWER

6-26

For this periodic voltage: a) Find the average value b) Find the effective value c) Find the average power dissipated in a 5 ohm resistor.

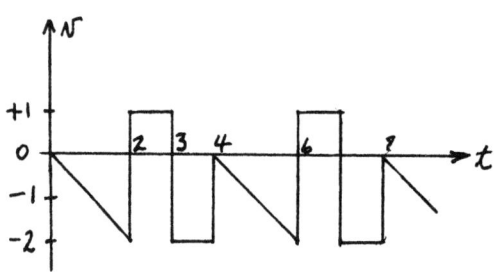

****

a)
$$V_{AVG} = \frac{1}{T}\int_0^T v(t)\,dt = \frac{1}{4}\left[\frac{1}{2}(2)(-2) + 1(1) + 1(-2)\right] = -\frac{3}{4}\ V.$$

b)
$$V_{RMS}^2 = \frac{1}{T}\int_0^T v^2(t)\,dt = \frac{1}{4}\left[\int_0^2 (-t)^2\,dt + 1(1) + (-2)^2\cdot 1\right]$$

$$= \frac{1}{4}\left[\frac{t^3}{3}\Big|_0^2 + 1 + 4\right] = \frac{1}{4}\left[\frac{8}{3} + \frac{15}{3}\right] = \frac{23}{12}$$

$$\therefore V_{RMS} = \sqrt{\frac{23}{12}} = 1.38\ V.$$

c)
$$AVG.\ POWER = \frac{V_{RMS}^2}{R} = \frac{23/12}{5} = .38\ WATT$$

AC Steady State Power / 195

6-27

Two loads are in parallel across 2400-V rms lines. One load absorbs 20 kW at a 0.9 lagging power factor, and the other load absorbs 30 kW at a 0.75 leading power factor. Find the total rms line current to the two loads.

***

The best approach is to obtain the apparent power from the magnitude of the total complex power, and then divide this apparent power by the line voltage. The resulting quotient is the desired total line current. Of course, the total complex power is the sum of the individual complex powers.

From power triangle considerations, each individual complex power has a magnitude that is equal to the real power divided by the power factor, and an angle that has a magnitude equal to the arc-cosine of the power factor, and that is positive for a lagging power factor but that is negative for a leading power factor. So, the total complex power S is

$$S = \frac{20\,000}{0.9} \angle \cos^{-1} 0.9 + \frac{30\,000}{0.75} \angle -\cos^{-1} 0.75$$

$$= 22\,222 \angle 25.8° + 40\,000 \angle -41.4°$$

$$= (20\,000 + j9686) + (30\,000 - j26\,457)$$

$$= 50\,000 - j16\,771 = 52\,738 \angle -18.5° \text{ V·A}$$

Finally, $I_L = \frac{52\,738}{2400} = 21.97$ A rms

6-28

A source with a complex internal impedance is connected to a load as shown. The load draws 1 kw of average power at 100 V rms with a power factor of 0.80 lagging. Determine the source voltage $v_s(t)$ and the type and value of the element to be placed in parallel with the load so that maximum power is transferred to the load. The radian frequency is 200 rad/s.

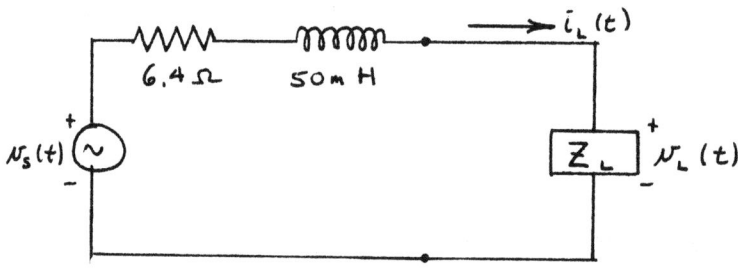

****

The complex power is:

$$\dot{P}_L = 1 + j \tan[\cos^{-1}(PF)] = 1 + j \tan[\cos^{-1}(.8)]$$

$$= 1 + j\,7.5 \text{ kVA} = 1000 + j\,750 \text{ VA}$$

Let $V_L = 100 \angle 0° \text{ V}$, then since $\dot{P}_L = V_L I_L^*$

$$I_L^* = 10 + j\,7.5, \quad I_L = 10 - j\,7.5 \text{ A}$$

$$Z_L = V_L / I_L = \frac{100}{10 - j\,7.5} = \frac{100(10 + j\,7.5)}{100 + 56.25}$$

$$= 6.4 + j\,4.8$$

$$V_s = V_L + I_L Z_s = 100 + (10 - j\,7.5)(6.4 + j\,200 \times 50 \times 10^{-3})$$

$$V_s = 100 + 64 - j\,48 + j\,100 + 75$$

$$= 239 + j\,52 = \sqrt{(239)^2 + (52)^2} \angle \tan^{-1} \frac{52}{239}$$

$$= 244.6 \angle 12.27° \text{ V}$$

$$v_s(t) = 244.6 \cos(200t + 12.27°) \text{ V}$$

For maximum power transfer, $X_L = -X_S$ and $R_L = R_S$. R_L already equals R_S so we need $X_L = -10$. If an unknown is placed in parallel with the load, the unknown reactance X_u combines with the 4.8 Ω reactance as:

$$\frac{j4.8(jX_u)}{j4.8 + jX_u} = jX_L = -j10$$

$$-4.8X_u = 48 + 10X_u$$

$$-14.8X_u = 48$$

$$X_u = -3.24 = -\frac{1}{\omega C}$$

$$C = -\frac{1}{\omega X_u} = \frac{1}{3.24 \times 200} = 1543 \mu F$$

We must place a 1543 μF capacitor in parallel with the load in order to transfer maximum power.

6-29

A fully loaded single-phase 10-hp (output rating) induction motor is connected across 240-V rms, 60-Hz lines. The motor operates at 80% efficiency and a 0.7 lagging power factor. A capacitor is to be connected across the input lines to obtain an overall power factor of 0.85 lagging. Determine the capacitance of the capacitor.

****

The input power to the motor is

$$\frac{10(745.7)}{0.80} = 9321 \text{ W}$$

which will remain the same with the capacitor connected in parallel. The reactive power that the capacitor must supply is equal to the initial reactive power minus the final reactive power. From power triangle considerations this difference ΔQ in reactive power is

$$\Delta Q = P \tan \theta_i - P \tan \theta_f$$

in which θ_i and θ_f are the initial and final power factor angles. These are, from the specified power factors,

$$\theta_i = \cos^{-1} 0.7 = 45.57°$$
$$\theta_f = \cos^{-1} 0.85 = 31.79°$$

So, $\Delta Q = 9321 (\tan 45.57° - \tan 31.79°) = 3733$ VAR

But, $\Delta Q = \omega C V^2$. So, $C = \dfrac{\Delta Q}{\omega V^2} = \dfrac{3733}{2\pi(60)(240)^2} = 172 \mu F$

6-30

If the voltage waveform shown is fed to a resistor immersed in water, it will heat the water as much as what value of dc voltage?

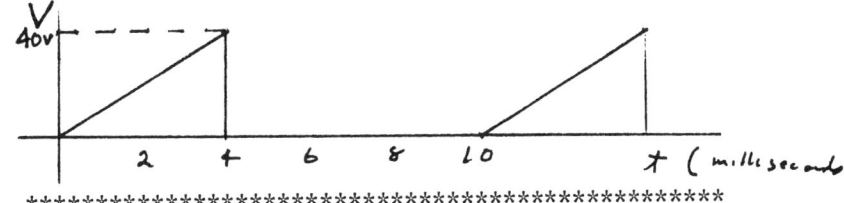

*****

For a non-dc voltage, the "effective value" is the "root-mean-square" (rms) value. It is given by

$$V_{eff} = \sqrt{\frac{1}{T}\int_T v^2(t)\,dt}$$

Now, $v(t) = mt$ between $t=0$ and $t=4\times 10^{-3}$. We can evaluate "m" by saying $40v = m(4\times 10^{-3})$; $m = 10^4$

$$V_{eff} = \sqrt{\frac{1}{10\times 10^{-3}}\int_0^{4\times 10^{-3}}(10^4 t)^2\,dt + \int_{4\times 10^{-3}}^{10^{-2}} 0\,dt}$$

$$= \sqrt{100\times 10^8 \int_0^{4\times 10^{-3}} t^2\,dt} = \sqrt{10^{10}\cdot \frac{t^3}{3}\Big]_0^{4\times 10^{-3}}} = \sqrt{\frac{10^{10}}{3}\times 64\times 10^{-9}}$$

$$= \sqrt{640/3} = \boxed{14.6\ v}$$

6-31

For the periodic waveform shown, find the average value and the rms value.

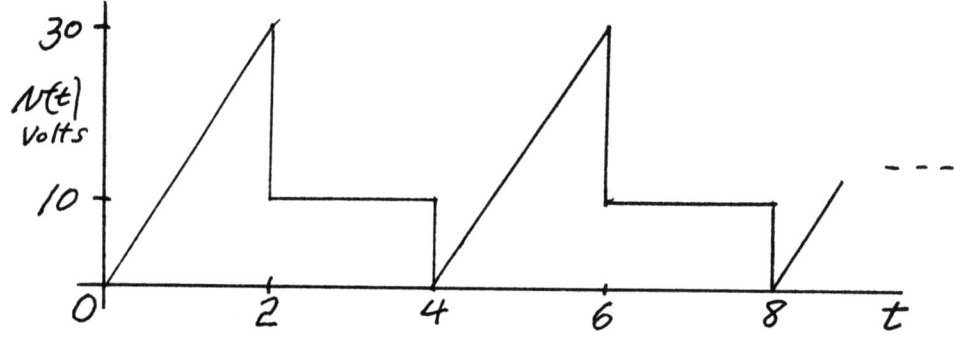

****

$$V_{AV} = \frac{1}{4}\int_0^4 v(t)\,dt = \frac{\text{area under one cycle}}{4}$$

$$V_{AV} = \frac{\frac{1}{2}(30)(2) + 10(2)}{4} = \frac{50}{4} = 12.5 \text{ volts}$$

$$V_{ms} = \frac{1}{4}\left[\int_0^2 (15t)^2\,dt + \int_2^4 10^2\,dt\right]$$

$$= \frac{1}{4}\left[\frac{225\,t^3}{3}\bigg|_0^2 + 100t\bigg|_2^4\right]$$

$$= \frac{1}{4}\left[600 + 200\right]$$

$$V_{ms} = 200$$

$$V_{rms} = \sqrt{200} = 14.142 \text{ volts}$$

6-32

Let $Z_L = 10 + j3\,\Omega$ in the motor drive circuit shown in the figure below, then calculate the average power delivered to the motor, the power factor of the motor drive source, and the average power delivered by the source.

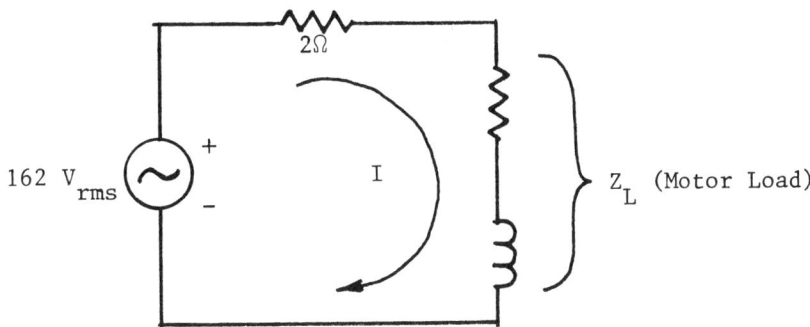

***

$$I_{rms} = V_{rms}/Z = 162/(2+10+j3) = 162/(12+j3)$$
$$= 54/(4+j1) = 54/(4.123\,\angle 14.03°)$$
$$= 13.096\,\angle -14.03°\ A_{rms}$$

Power Factor $= P_f = \cos\theta_{VI} = \cos(-14.03°) = 0.9701$

The Average Power Can Be Calculated As

$$P_{Del.\,Source} = V_s \cdot I \cdot P_f = 162 \times 13.096 \times 0.9701 = 2058.1\,W.$$

$$P_{Load} = I^2 R_m = (13.096)^2 \cdot 10 = 1715.1\,W.$$

Check:

$$P_{Del.\,Source} = P_{Load} + P_{2\Omega}$$

$$2058.1 = 1715.1 + (13.096)^2 \cdot 2 = 1715.1 + 343$$
$$2058.1 = 2058.1$$

6-33

The load draws 33 kw at a power factor of 0.75 lagging.
a) Find the value of C to raise the power factor of the circuit to 0.9 lagging.
b) Find the value of I_{rms} before and after C is added.

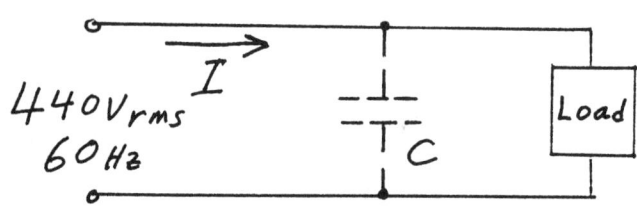

a)

Using the idea of complex power we have:

For a pf of 0.75:

$P = 44$ KVA, $Q = 29.1$ KVAR, angle $41.4°$, $P = 33$ kw

For a pf of 0.9:

$P = 36.67$, $Q = 15.98$, angle $25.8°$, $P = 33$ kw

Therefore we need $Q_c = 15.98 - 29 = -13.12$ KVAR

$Q_c = -V^2 \omega C$

$C = \dfrac{13,120}{440^2 \times 377} = 179.77 \mu F$

b)

before C added:

$440 \, I_{rms}^2 (.75) = 33 \times 10^3$

$I_{rms} = 100 A$

after C added:

$440 \, I_{rms}^2 (.9) = 33 \times 10^3$

$I_{rms} = 83.33 A$

6-34

For the two impedances in parallel and the additional phasor information given, calculate for impedance 1 its complex power, power factor, and terminal current. Then calculate for impedance 2 its terminal current, complex power, and power factor. Last calculate the total complex power and power factor for the total circuit. (Voltages and currents are given in effective values.)

$\vec{V}_S = 100 \angle 160°$ V.

$\vec{I}_S = 2 \angle 190°$ A.

$P_1 = 23.2$ W.

$Q_1 = +50$ VARS

****

$\vec{S}_1 = P_1 + j Q_1 = 23.2 + j50 = \underline{55.12 \angle 65.1° \text{ VA.}}$

$pf_1 = \cos(65.1°) = \underline{0.42 \text{ lag}}$

$\vec{I}_1 = \left(\dfrac{\vec{S}_1}{\vec{V}_S}\right)^* = \left(\dfrac{55.12 \angle 65.1°}{100 \angle 160°}\right)^* = \underline{0.551 \angle 94.9°} = -0.047 + j0.549$ A.

$\vec{I}_2 = \vec{I}_S - \vec{I}_1 = -1.97 - j0.34 + 0.047 - j0.549 = \underline{2.12 \angle -155° \text{ A.}}$

$\vec{S}_2 = \vec{V}_S \vec{I}_2^* = (100 \angle 160°)(2.12 \angle 155°) = 212 \angle -45° = \underline{150 - j150 \text{ VA.}}$

$pf_2 = \cos(-45°) = \underline{0.707 \text{ lead}}$

$\vec{S}_T = \vec{S}_1 + \vec{S}_2 = 173.2 - j100 = \underline{200 \angle -30° \text{ VA.}}$

$pf_T = \cos(-30°) = \underline{0.866 \text{ lead}}$

6-35

Determine the real power consumed by the load. Use the mesh analysis method.

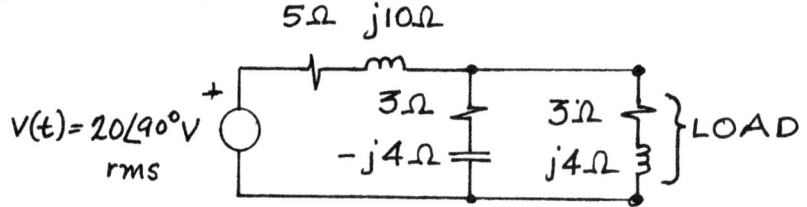

****

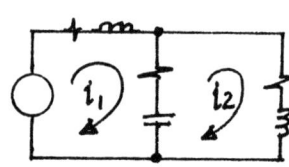

ASSUMED CURRENTS

KVL EQS:

$$i_1(8+j6) - i_2(3-j4) = 20\angle 90°$$

$$-i_1(3-j4) + i_2 \cdot 6 = 0$$

$$i_1 = i_2\left(\frac{6}{3-j4}\right) \quad \therefore \quad (8+j6)\left(\frac{6}{3-j4}\right)i_2 - (3-j4)i_2 = 20\angle 90°$$

$$-i_2(3-j16) = 20\angle 90°$$

$$i_2 = 1.23 \angle -10°$$

REAL POWER $= i_2^2 R_L = (1.23)^2 \times 3 = 4.54\,W$

6-36

a.) In the circuit shown below find the complex power, $\bar{P}$.
b.) What is the power factor, p.f.
c.) What size capacitor in parallel with the load will make the composite p.f. = 0.95.

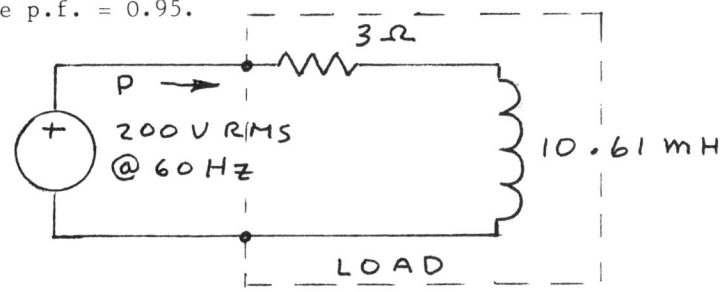

****

a) $\bar{Z} = R + j\omega L = 3 + j\, 2\pi 60 \cdot 10.61 \cdot 10^{-3}$
$= 3 + j4 = 5\, \underline{/53.1°}\ \Omega$

$\bar{I} = \dfrac{\bar{V}}{\bar{Z}} = \dfrac{200\, \underline{/0}}{5\, \underline{/53.1°}} = 40\, \underline{/-53.1°}$ A.

$\bar{P} = \bar{V}\bar{I}^* = 200 \cdot 40\, \underline{/53.1°}$
$= 8000\, \underline{/53.1°}$ VA   ANS

b) P.F. = $\cos\theta$ = $\cos 53.1°$ = 0.6   ANS

c) POWER PHASOR DIAGRAM:
$\bar{P} = 4800 + j6400$ VA

REACTIVE POWER TO BE ADDED BY CAPACITOR = P_c
DESIRED POWER
$\cos^{-1}.95$

$P_c = 6400 - \dfrac{4800}{.95}\sqrt{1-.95^2} = 4820$ VAR

$P_c = \dfrac{V^2}{X_c}$

$4820 = 200^2 \cdot 2\pi 60\, C$

$C = 320\ \mu F$ ANS

6-37

For the circuit shown, with a time dependent voltage source of v(t) = 65 sin 377t V, determine the power consumed by the 16 Ω resistance.

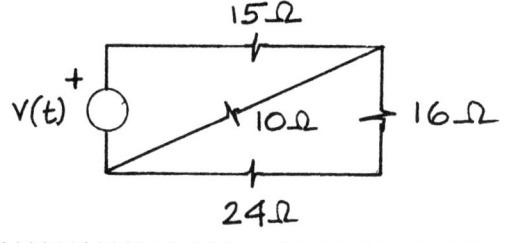

PEAK VOLTAGE IS 65V FOR SOURCE

$V_{rms} = \frac{65}{\sqrt{2}} = 45.96V$ USED FOR POWER CALCULATIONS

FIND EQUIVALENT RESISTOR FOR 10, 16, 24 Ω COMBINATION AND APPLY VOLTAGE DIVIDER RULE.

$R_e = \frac{10 \times (16+24)}{10+16+24} = 8\Omega$

$V_{8\Omega} = V_{rms} \times \frac{8}{8+15} = 45.96 \times \frac{8}{23} = 16V$

$V_{16\Omega} = V_{8\Omega} \times \frac{16}{16+24} = 16 \times \frac{16}{40} = 6.40V$

$POWER = \frac{(V_{16\Omega})^2}{R_{16}} = \frac{(6.40)^2}{16} = 2.56W$

6-38

A series circuit connected to a 120 V, 60 Hz supply draws 120 W at 0.8 power factor leading. Determine the series equivalent of the circuit. Develop the parallel equivalent of the circuit.

****

Power = Volts × amps. × power factor = $VI \cos\theta$

$I = \dfrac{P}{V\cos\theta} = \dfrac{120}{120 \times 0.8} = 1.25\text{ A}$; $|Z| = \dfrac{120\text{ V}}{1.25\text{ A}} = 96\,\Omega$

In rectangular form: $Z = R - jX_c = 96(\cos\theta - j\sin\theta)$
$= 96(0.8 - j\,0.6) = 76.8 - j\,57.6\,\Omega$

Hence $R = 76.8\,\Omega$ and Capacitive reactance, $X_c = 57.6\,\Omega$

$X_c = \dfrac{1}{\omega C} = \dfrac{1}{2\pi f C}$. So $C = \dfrac{1}{2\pi f X_c} = \dfrac{1}{2\pi \times 60 \times 57.6}$ F
$= 46.05\,\mu F$

The series circuit is therefore of the form shown

To obtain the parallel equivalent circuit the admittance function is obtained.

$Y = \dfrac{1}{Z} = \dfrac{1}{R - jX_c} = \dfrac{R + jX_c}{(R - jX_c)(R + jX_c)} = \dfrac{R + jX_c}{R^2 + X_c^2} = \dfrac{R + jX_c}{|Z|^2}$

So $Y = \dfrac{R}{|Z|^2} + j\dfrac{X_c}{|Z|^2} = G + jB$

For the parallel circuit shown,

$R_p = \dfrac{1}{G} = \dfrac{|Z|^2}{R} = \dfrac{96^2}{76.8} = 120\,\Omega$

$X_{c_p} = \dfrac{1}{B} = \dfrac{|Z|^2}{X_c} = \dfrac{96^2}{57.6} = 160\,\Omega$

Now $X_{c_p} = \dfrac{1}{\omega C_p}$. So $C_p = \dfrac{1}{\omega X_{c_p}} = \dfrac{1}{2\pi f \cdot X_{c_p}}$

Hence $C_p = \dfrac{1}{2\pi \times 60 \times 160}$ F $= \dfrac{10^6}{2\pi \times 60 \times 160}\,\mu F = 16.58\,\mu F$

6-39

Two loads are connected in parallel across a 2000 V(RMS) 60 Hz line. One load absorbs 1000 kW at a 0.5 power factor lagging. The second load absorbs 600 kW and 800 kVAR. A capacitor is to be added in parallel to the two loads so as to improve the overall p.f. to 0.85 lagging.
(a) Specify the value of the capacitor.
(b) What is the RMS magnitude of the line current after the capacitor has been added?

****

a) FOR LOAD 1, $\underline{S}_1 = P_1 + jQ_1 = (1000 + j1732)\,kVA$

FOR LOAD 2, $\underline{S}_2 = P_2 + jQ_2 = (600 + j800)\,kVA$

COMBINED, $\underline{S}_1 + \underline{S}_2 = (1600 + j2532)\,kVA$

$\cos\theta_{Tot} = 0.85 \Rightarrow \theta_{Tot} = 31.8° \Rightarrow \tan\theta = 0.62 = \frac{Q}{P} = \frac{Q}{1600} \Rightarrow Q = 992\,kVAR$

A p.f. of 0.85 LAG REQUIRES 992 kVARS FOR A POWER OF 1600 kW. HENCE, C MUST ABSORB 992 kVAR − 2532 kVAR = −1540 kVARS.

SINCE $Q = |V_{RMS}|^2 / X = -\omega C |V_{RMS}|^2$

THEN, $C = \frac{Q}{-\omega|V_{RMS}|^2} = \frac{-1540 \times 10^3}{-377(2000)^2} = 1021\,\mu F$

b) AFTER C HAS BEEN ADDED,

$\underline{S}_T = (1600 + j992)\,kVA = 1882\,\underline{/31.8°}\,kVA$

THEREFORE, $|I| = \frac{|\underline{S}_T|}{|V_{RMS}|} = \frac{1882 \times 10^3}{2 \times 10^3} = 941\,A.$

6-40

If 250 watts is dissipated in the resistance, find (a) V, (b) the source current, (c) circuit power factor, (d) magnitude of source voltage.

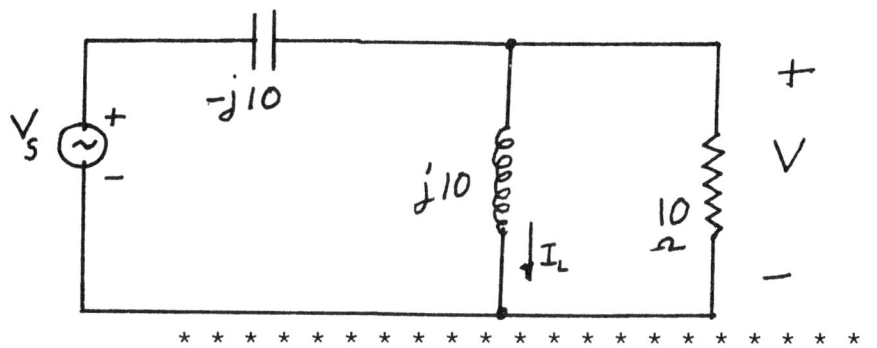

(a) $P = I^2 R$, $\quad I^2 = \dfrac{250}{10} = 25$

$|I| = 5$ amps.   Choose $0°$ reference for current thru resistor.

$V = 50 \angle 0°$

(b) $I_L = \dfrac{50 \angle 0°}{10 \angle 90°} = 5 \angle -90° = -5j$

$I_{SOURCE} = I_R + I_L = 5 - 5j = 5\sqrt{2} \angle -45°$

(c) Circuit $Z = -10j + \dfrac{10(10j)}{10 + 10j} = -10j + 5 + 5j$

$= 5 - 5j = 5\sqrt{2} \angle -45°$

Circuit $PF = \cos 45° = .7071$, lagging

(d) Since all quantities are rms,

$P = V_S I_S \cos \phi$

$|V_S| = \dfrac{P}{|I_S| \cos \phi} = \dfrac{250}{5\sqrt{2} \cos 45°} = 50$ Volts

6-41

Find the average and rms value of the voltage waveform.

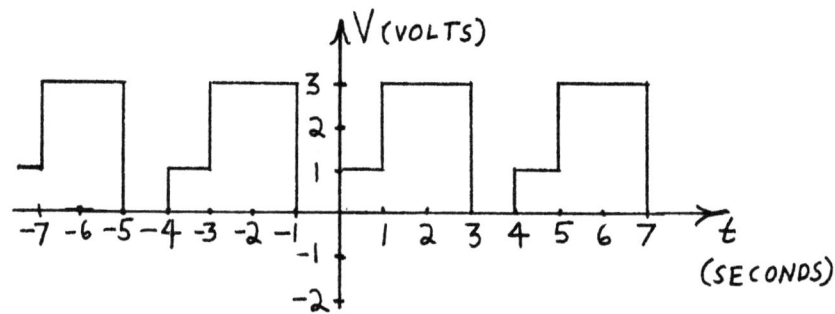

$T = 4$ seconds (TIME OF ONE CYCLE)

$$V_{AVERAGE} = \frac{\text{AREA UNDER ONE CYCLE}}{\text{TIME OF ONE CYCLE}}$$

$$= \frac{1 \times 1 + 3 \times 2 + 0 \times 1}{4} = \underline{1.75 \text{ VOLTS}}$$

$$V_{RMS} = \left(\frac{\text{AREA UNDER ONE } V^2 \text{ CYCLE}}{\text{TIME OF ONE CYCLE}}\right)^{1/2}$$

$$= \left(\frac{1^2 \times 1 + 3^2 \times 2 + 0^2 \times 1}{4}\right)^{1/2} = \underline{2.18 \text{ VOLTS}}$$

6-42

Use two different methods to find C such that the 100 volt 60 Hz source sees a unity power factor (pf).

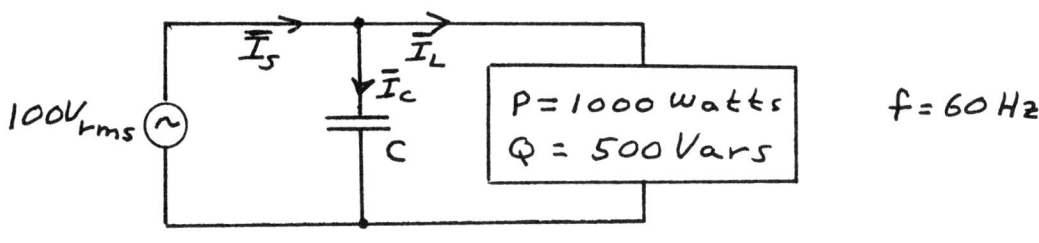

****

Method #1

Source complex power = $P_L + jQ_L$

Load complex power = $1000 + j\left(500 - \frac{|V|^2}{|X_c|}\right)$

For unity power factor $Q_L = 0$

$$\frac{|V|^2}{|X_c|} = 500 = (100)^2 \omega C$$

$$C = \frac{500}{100^2 \times 2\pi \times 60} = \boxed{132.6 \mu f}$$

Method #2

Assume input voltage = $100\angle 0$

$\overline{V}\overline{I_L}^* = (100\angle 0)(\overline{I_L}^*) = 1000 + j500$

$\overline{I_L} = 10 - j5$

For unity power factor $\angle \overline{V} = \angle \overline{I}$

$\angle \overline{I_s} = 0$

$\overline{I_s} = \overline{I_c} + \overline{I_L} = -j\omega C(100) + 10 + j5$

$= 10 + j0$

$100 \omega C = 5 \longrightarrow \boxed{C = 132.6 \mu f}$

6-43

Find the average value, the average magnitude (absolute value) and the rms value for the repetitive current which has one period defined as shown.

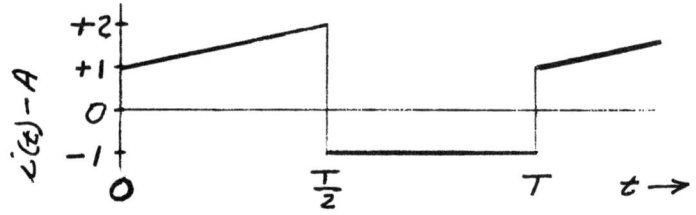

Define $i(t)$:
$$i(t) = \left(\frac{2}{T}\right)t + 1 \qquad 0 < t < \frac{T}{2}$$
$$i(t) = -1 \qquad \frac{T}{2} < t < T$$

Average Value $= \frac{1}{T}\left[\int_0^{T/2}\left(\frac{2t}{T}+1\right)dt + \int_{T/2}^{T}(-1)dt\right]$

$= \frac{1}{T}\left[\frac{T^2}{4T} + \frac{T}{2} - T + \frac{T}{2}\right] = \frac{1}{4} = 0.25$ A

Average Magnitude $= \frac{1}{T}\left[\int_0^{T/2}\left(\frac{2t}{T}+1\right)dt + \int_{T/2}^{T}(+1)dt\right] = 1.25$ A

rms Value $= \sqrt{\frac{1}{T}\left\{\int_0^{T/2}\left(\frac{2t}{T}+1\right)^2 dt + \int_{T/2}^{T}(-1)^2 dt\right\}}$

rms Value $= \sqrt{\frac{10}{6}} = 1.2910$ A

THREE PHASE CIRCUITS

━━━━━━━━━━━━━━━━━━━━━━━━━━━━━━━━━━━6-44

Find the total Watts, Vars and Volt-amperes used by the load in the balanced three phase circuit. Also find the power factor of the load.

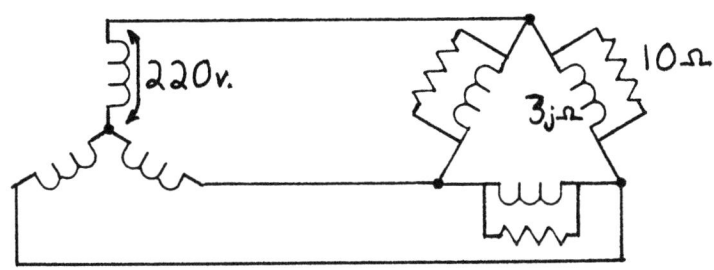

****

$V_L = 220\sqrt{3} = 381$ VOLTS

$P_\phi = V_\phi^2/R_\phi = V_L^2/R_\phi = 381^2/10 = 14.5$ KW

$P_T = 3P_\phi = 3(14520) = \underline{43.6\text{ KW}}$

$Q_\phi = V_\phi^2/X_{L\phi} = 381^2/3 = 48.4$ KVAR

$Q_T = 3Q_\phi = 3(48,400) = \underline{145.2\text{ KVAR}}$

$S_T = (P_T^2 + Q_T^2)^{1/2} = \underline{151.6\text{ KVA}}$

POWER FACTOR $= \dfrac{P_T}{S_T} = \dfrac{43.6}{151.6} = \underline{.288\text{ LAGGING}}$

6-45

The magnitude of the line voltage at the terminals of a balanced Y-connected load is 240 V(RMS). The load impedance is 10 $\underline{/-30°}$ ohms/phase. The load is fed from a line that has an impedance of (0.8 + j 1.0) ohms/phase. (a) Construct a single phase equivalent circuit. (b) What is the RMS magnitude of the line current? (c) What is the RMS magnitude of the line voltage at the generator?

****

a)

[Circuit diagram: Source $\bar{V}_{an}$ between nodes a (top, +) and n (bottom, −), connected through a 0.8 Ω resistor and a j1 Ω inductor to node A. Between A and N is the load: 240/√3 V with 8.66 Ω in series with −j5 Ω.]

b) $\bar{I}_{LINE} = \dfrac{240/\sqrt{3}}{10} = 13.86$ A.

c) $\bar{V}_{an} = \dfrac{240}{\sqrt{3}} + \dfrac{24}{\sqrt{3}} \underline{/30°}\,(0.8+j1)$

$= 142.32 \underline{/7.08°}$ V.

$V_{LINE} = 142.32\,(\sqrt{3}) = 246.5$ VOLTS

6-46

A balanced three phase delta load each consisting of a 20 ohm resistance in series with a 0.053 Henry inductor is supplied by a 60 cps, three phase generator, with 440 volts line-to-line.
Determine the line current, and the real reactive power consumed by the load.

****

$X_L = 2\pi f \times 0.053 = 20 \, \Omega$

PHASE LOAD $Z_p = 20 + j20 \, \Omega$

PHASE CURRENT FOR Δ CONNECTION, $I_p = \dfrac{V_L}{Z_p}$

$I_p = \dfrac{440 \angle 0°}{28.3 \angle 45°} = 15.6 \angle -45° \, A$

LINE CURRENT FOR Δ, $I_L = \sqrt{3} \, I_p = 26.9 \, A$

REAL POWER $= 3 I_p^2 R = 3 \times 15.6^2 \times 20 = 14.6 \, kW$

REACTIVE POWER $= 3 I_p^2 X = 14.6 \, kvar$

6-47

A balanced 3-phase wye connected load consists of a 15 ohm resistor and a 0.02 H inductor in series for each phase. This load is connected to a 3-phase 60 Hz power line with a 240 V (rms) line voltage. Calculate the total power dissipated in the load and the rms value of the line currents.

****

Phase impedance: $Z_{phase} = 15 + j(120\pi)(0.02) = 16.788 \angle 0.4658$

Phase voltage: $V_{phase} = \dfrac{240}{\sqrt{3}} = 138.564 \, V$

Line Current = phase Current = $\dfrac{138.564}{16.788 \angle 0.4658} = 8.254 \angle 0.4658$

Phase power = $(138.564)(8.254) \cos 0.4658 = 1021.86 \, W$

Total Power = 3 (Phase power) = 3 (1021.86) = 3065.58 W

6-48

A delta-connected 3-phase load absorbs 30 kw at a leading power factor of 0.8 when the line-to-line voltage is 4000 v rms

A. Find the phase impedances.

B. What must be the line-to-line voltage at the generator end of the line if each line contains impedance of 1 + j5 ohms?

****

A. Assuming the author meant to say the load is balanced, each phase absorbs $30 \text{Kw}/3 = 10 \text{Kw}$

$10,000 = |V_p||I_p| \text{ P.F} = 4000 |I_p| \times 0.8 = 3200 |I_p|$

$|I_p| = \frac{10,000}{3,200} = 3.125 \text{ A} \qquad |Z_p| = \frac{|V_p|}{|I_p|} = \frac{4000}{3.125} = 1280 \, \Omega$

$R = |Z_p| \times \text{P.F.} = 1280 \times 0.8 = 1024 \, \Omega$

$|X| = |Z_p|\sqrt{1-(\text{P.F.})^2} = 1280 \times \sqrt{1-.64} = 1280 \sqrt{.36}$

$= 768$. For leading power factor, this means

$$\boxed{Z_p = 1024 - j768}$$

B. We need to convert the delta-connected load to an equivalent wye: $Z_{p_Y} = Z_{p_\Delta}/3 = 341.3 - j256$

We may say we have $4000/\sqrt{3}$ across Z_{p_Y} and we want the total voltage across the series connection of Z_{p_Y} and line impedance

$\dfrac{V_{\text{Line to ground}}}{341.3 - j256 + 1 + j5} = \dfrac{4000/\sqrt{3}}{341.3 - j256}$; Line-to-line

voltage is $\dfrac{4000}{|341.3 - j256|} |342.3 - j251| = \dfrac{4000}{426.6} \times 424.5$

$= \boxed{3979.6 \text{ v}}$

6-49

For the positive sequence, balanced 3 phase circuit and the additional information given, calculate the phase voltage from node a to node n, the line-to-line voltage from node b to node c, the power factor of the circuit and the total complex power.

$\vec{V}_{bn} = 100 \angle 0°$ V.
$\vec{Z}_Y = 5 - j6\ \Omega$

***

POSITIVE SEQUENCE $\Rightarrow \vec{V}_{an}$ LEADS $\vec{V}_{bn}$ BY $120° \Rightarrow \underline{\vec{V}_{an} = 100 \angle 120°}$

$\vec{I}_L = \dfrac{\vec{V}_{an}}{\vec{Z}_Y}$ & $\vec{Z}_Y = 5 - j6 = 7.81 \angle -50.2°\ \Omega$

$\vec{I}_L = \dfrac{100 \angle 120°}{7.81 \angle -50.2°} = \underline{12.8 \angle 170.2°\ A.}$

$\vec{V}_{bc} = \vec{V}_{bn} + \vec{V}_{nc} = (\sqrt{3})\vec{V}_{bn} \angle 30° = (1.732)(100 \angle 0°) \angle 30°$

$\Rightarrow \underline{\vec{V}_{bc} = 173.2 \angle 30°\ V.}$

$pf = \cos(-50.2°) = \underline{0.64\ \text{lead}}$

$\vec{S}_\phi = \vec{V}_\phi \vec{I}_\phi^* = \vec{V}_{an} \vec{I}_L^* = (100 \angle 120°)(12.8 \angle -170.2°) = \underline{1280 \angle -50.2°\ VA.}$

$\vec{S}_T = 3\vec{S}_\phi = \underline{3840 \angle -50.2°\ VA.}$

FREQUENCY RESPONSE

6-50

Calculate the resonant frequency, the Q of the circuit and the half-power bandwidth for the circuit shown and the given parameter values. Assume that Q is large.

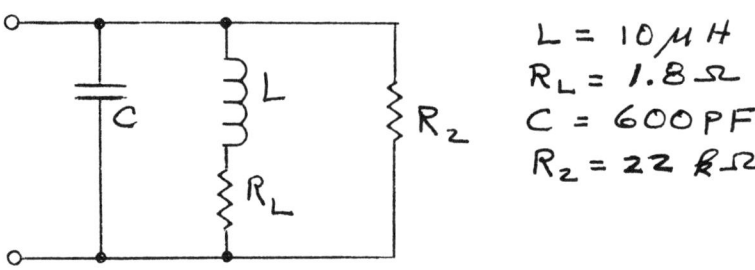

$L = 10 \mu H$
$R_L = 1.8 \, \Omega$
$C = 600 pF$
$R_2 = 22 \, k\Omega$

****

Assuming Q is high, $\omega_0 = \dfrac{1}{\sqrt{LC}} = 12.909$ Mrad/s

$f_0 = 2.0547$ MHz

$\omega_0 L = (12.909 \times 10^6)(10 \times 10^{-6}) = 129.09$

Inductor branch: $Y_L = \dfrac{1}{1.8 + j129.09} = 0.00774578 \underline{/-1.556854}$

$Re(Y_L) = 0.00010799$ S ; $G_2 = \dfrac{1}{22,000} = 0.000045454$

Effective shunt $R = \dfrac{1}{Re(Y_L) + G_2} = 6517.036 \, \Omega$

$Q_0 = \omega_0 C R_{shunt} = (12.909 \times 10^6)(600 \times 10^{-12})(6517.036)$

$Q_0 = 50.48$

$\Delta f = \dfrac{f_0}{Q_0} = \dfrac{2.0547 \times 10^6}{50.48} = 40,703.25$ Hz

6-51

Plot the magnitude of H vs frequency and the phase angle of H vs frequency for the circuit shown.

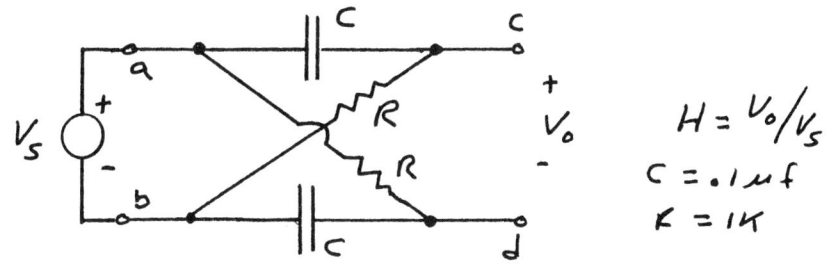

$H = V_o/V_s$
$C = .1 \mu f$
$R = 1K$

***

$V_d = V_s \dfrac{1/sc}{R + 1/sc}$

$= V_s \dfrac{1}{sCR + 1}$

$V_c = V_s \dfrac{R}{R + 1/sc}$

$= V_s \dfrac{sCR}{sCR + 1}$

$V_o = V_c - V_d$

$= V_s \dfrac{sCR - 1}{sCR + 1}$

$H(j\omega) = \dfrac{V_o}{V_s}(j\omega) = \dfrac{j\omega RC - 1}{j\omega RC + 1} = -\dfrac{1 - j\omega RC}{1 + j\omega RC}$

$\left|\dfrac{V_o}{V_s}\right| = \dfrac{\sqrt{\omega^2 R^2 C^2 + 1}}{\sqrt{\omega^2 R^2 C^2 + 1}} = 1$

$\angle \dfrac{V_o}{V_s} = 180 - \tan^{-1}\omega RC - \tan^{-1}\omega RC$

$= 180 - 2\tan^{-1}\omega RC$

$= 180 - 2\tan^{-1}\dfrac{\omega}{10^4}$

6-52

For the transfer function T(s) shown, find:
  A. The steady state magnitude of T(s), s = jω.
  B. The steady state phase of T(s), s = jω.
  C. What is the phase angle at ω = 2rps?

$$\frac{V_o}{V_i} = T(s) = \frac{2s^2 - 3s + 4}{3s^3 + 4s^2 + 5s + 6}$$

****

A.

$$T(s)\Big]_{s=j\omega} = \frac{-2\omega^2 + j(-3\omega) + 4}{-4\omega^2 + 6 + j(5\omega - 3\omega^3)}$$

$$|T(s)| = \sqrt{\frac{(4-2\omega^2)^2 + (-3\omega)^2}{(6-4\omega^2)^2 + (5\omega - 3\omega^3)^2}}$$

B.

$$\angle T(s) = \tan^{-1}\frac{-3\omega}{4-2\omega^2} - \tan^{-1}\frac{5\omega - 3\omega^3}{6-4\omega^2}$$

C. At ω = 2 rps

$$\angle T(s) = \tan^{-1}\frac{-6}{4-2(2)^2} - \tan^{-1}\frac{10 - 3(2)^3}{6 - 4(2)^2}$$

$$= \tan^{-1}\frac{-6}{-4} - \tan^{-1}\frac{-14}{-10}$$

(Both angles are in the 3rd quadrant)

$$\angle T(s) = 236.3° - 234.5° = 1.8°$$

6-53

Given the transfer function
$$H(s) = \frac{30\,000s(s + 0.5)}{(s + 2)^3}$$
for which a Bode magnitude plot is to be drawn. If the semilog paper used has the minimum number of cycles needed to include the break frequencies, then without drawing a Bode plot, analytically determine the values at the two ends of the Bode plot.

***

From the specified transfer function, the corner frequencies are 0.5 rad/s and 2 rad/s. So, a two-cycle semilog sheet is necessary with 0.1 rad/s being the lowest radian frequency and 10 rad/s being the greatest radian frequency. Since 0.1 rad/s is less than the lowest break frequency, the s in each s+T factor is considered negligible, giving an approximation of

$$|H(j0.1)| \cong \frac{30\,000|j0.1|(0.5)}{2^3} = 187.5$$

which in decibels is $20 \log 187.5 = 45.5\,dB$

At the other end of the Bode plot, the radian frequency of 10 rad/s is greater the greatest break frequency, and so at this frequency the approximation is that the T in each s+T factor is negligible. The result is

$$|H(j10)| \cong \frac{30\,000|j10||j10|}{|j10|^3} = 3000$$

which in decibels is $20 \log 3000 = 69.5\,dB$

Consequently, the Bode plot starts at 45.5 dB and ends at 69.5 dB.

6-54

Given the Bode magnitude plot below, find:
A. A simple pole-zero pattern that has this magnitude.
B. Find $T(s) = \dfrac{V_o}{V_i}$ for your pole-zero pattern.
C. What is the magnitude (in dB) of $T(s)$ as $\omega \to \infty$?

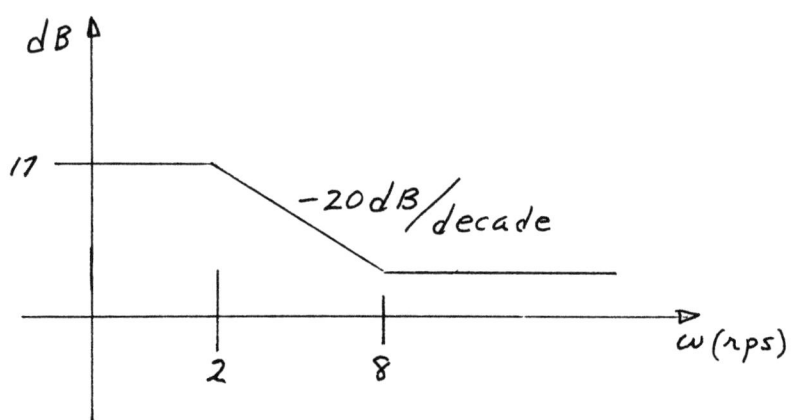

****

A. A single pole at -2 rps and a zero at -8 rps is the simplest pattern.

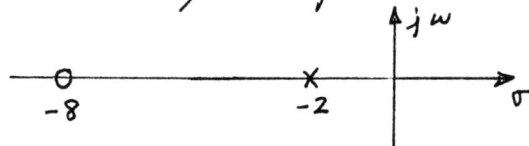

B.
$$T(s) = 1.77\left(\dfrac{s+8}{s+2}\right)$$

The multiplier is 1.77 since 17 dB was specified as $s \to 0$ that is $\omega \to 0$ and
$20 \log 1.77(4) = 17 \, dB$

C. Since $-20 \, dB/decade = -6 \, dB/octave$ and we have 2 octaves (2 to 8 rps), as $s \to \infty$ we are down 12 dB from the D.C. value.

as $\omega \to \infty$
$s \to \infty$ $|T(s)| = 17 - 12 = 5 \, dB$

6-55

Given the transfer function $H(s) = \dfrac{50,000}{(s+10)(s^2+10s+100)}$

A. Sketch a Bode amplitude response plot, showing gain in db for any flat portion, break frequency and slope of asymptotes.

B. If the input to the system is cos 10t, write the output in the form $A \cos(10t + \sigma)$, evaluating A and σ.

A

We divide to make the constants in each factor = 1

$$H(s) = \frac{50,000}{10 \times 100} \cdot \frac{1}{\left(1+\frac{s}{10}\right)\left(1+\frac{s}{10}+\frac{s^2}{100}\right)}$$

If we neglect the s term in the second factor, we would find a triple pole at $s = 10$. Each pole gives a slope of -20 db/decade. At zero and low frequency, $H(j\omega) = 50$ and $20 \log_{10} 50 \simeq 34$ db. Our Bode plot would be approximately:

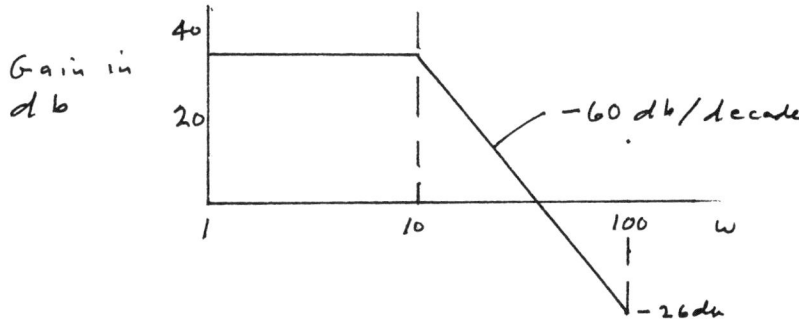

B.
$H(j\omega) = \dfrac{50,000}{(10+j10)(-10^2+j10\times10+100)} = \dfrac{50,000}{10\sqrt{2}\,\underline{/45°} \times 100\,\underline{/90°}}$

$= \dfrac{50}{\sqrt{2}}\,\underline{/-135°}$

Hence, output is $\boxed{\dfrac{50}{\sqrt{2}} \cos(10t - 135°)}$

6-56

In network analysis, we often are interested only in how a voltage or a current is transferred from an input source as shown in the figure given below.

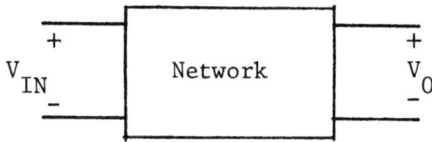

$V_0(\omega) = A \, V_{IN}(\omega)$ where A is the transfer function.

Given the following network, derive an expression for $A = V_0(\omega)/V_{IN}(\omega)$ using phasor analysis. Sketch the magnitude frequency response, i.e. $|A|$ versus ω.

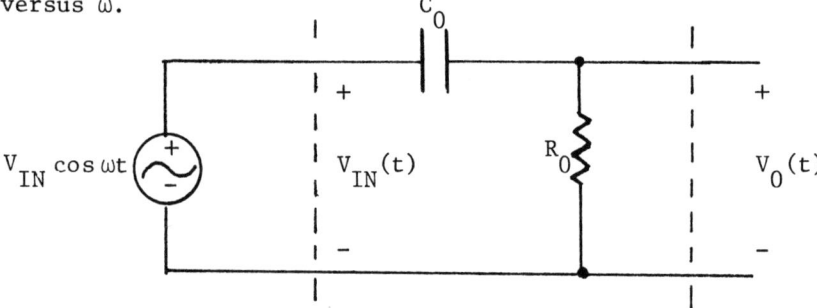

****

Using Phasor Representation, The Voltage Divider Law Gives

$$V_0(\omega) = \frac{R_0}{R_0 + \frac{1}{J\omega C_0}} V_{in}$$

Simplifying,

$$A(\omega) = \frac{V_0(\omega)}{V_{in}} = \frac{J\omega C_0 R_0}{1 + J\omega C_0 R_0}$$

Which becomes

$$|A(\omega)| = \left| \frac{J\omega C_0 R_0}{1 + J\omega C_0 R_0} \right| = \frac{\omega C_0 R_0}{\sqrt{1 + (\omega C_0 R_0)^2}}$$

For A Sketch Of $|A|$, The Main Points to be Considered Are

ω	$|A|$
0	0
$1/R_oC_o$	$1/\sqrt{2}$
∞	1

And the Sketch is shown Below.

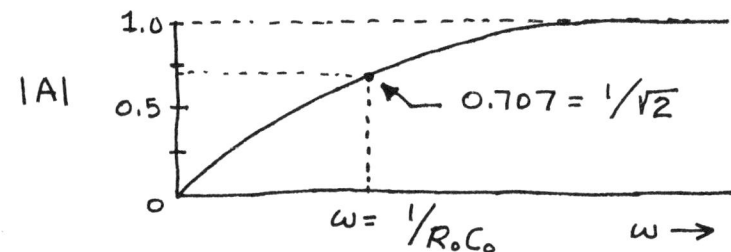

6-57

At what frequency will the signal source see a purely resistive load?

$R = 100\,\Omega$, $L = 15\,mh$, $C = 100\,pf$

*** *** *** **** ** * ** ** **** ** *** **** ** *** *** ** ******

At resonance; $f = \dfrac{1}{2\pi\sqrt{LC}}$

$= \dfrac{1}{2\pi\sqrt{15mh \cdot 100pf}} = 130\,khz$

6-58

Find R, L, C, and I_m, where $i_s(t) = I_m\cos\omega_0 t$ if the resonant frequency is 10 rad/s, the bandwidth is 2 rad/s, the magnitude of v at resonance is 100 volts and the current flowing in the inductor at resonance has a magnitude of 1.0A. Use frequency scaling to determine R, L, and C if the resonant frequency is to be 100 k rad/s.

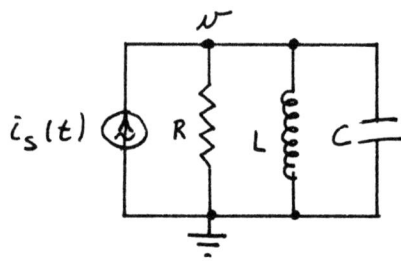

****

$$\omega_0 = 10 = \frac{1}{\sqrt{LC}} \qquad Q_0 = \frac{\omega_0}{B} = \frac{10}{2} = 5$$

$$Q_0 = 5 = \omega_0 RC = 10RC$$

In resonance, $v = i_s R$ so $V_m = I_m R = 100$

$$|I_L| = Q_0 I_m = 5 I_m = 1 \qquad I_m = 0.2 A$$

$$R = \frac{100}{I_m} = \frac{100}{0.2} = 500 \,\Omega$$

$$C = \frac{5}{10R} = \frac{5}{5000} = 10^{-3} F = 1000 \mu F$$

$$\sqrt{LC} = \frac{1}{10} \qquad LC = \frac{1}{100} \qquad L = \frac{1}{100C}$$

$$L = \frac{1}{0.1} = 10 H$$

To use frequency scaling: $K_f = \frac{100 \,k\,rad}{10 \,rad}$

$$K_f = \frac{10^5}{10} = 10^4 \qquad R_H = R = 500 \,\Omega \qquad L_H = \frac{L}{K_f}$$

$$L_H = \frac{10}{10^4} = 10^{-3} H = 1.0 mH$$

$$C_H = \frac{C}{K_f} = \frac{10^{-3}}{10^4} = 10^{-7} F = 0.1 \mu F$$

Frequency Response / 227

6-59

A certain one-port circuit has an applied voltage of
$$v(t) = 8 - 42 \cos(2t + 15°) + 75 \sin(5t - 40°) - 120 \cos(10t + 50°) \text{ V}$$
The impedance of the circuit has a magnitude frequency-response in ohms of
$$|Z| = \begin{cases} 4, & \omega = 0 \text{ rad/s} \\ 5\omega + 4, & 0 < \omega \le 4 \text{ rad/s} \\ 1.5\omega^2, & \omega > 4 \text{ rad/s} \end{cases}$$
and a phase frequency-response in degrees of
$$\underline{/Z} = \begin{cases} 0, & \omega = 0 \text{ rad/s} \\ 50, & 0 < \omega \le 4 \text{ rad/s} \\ 3\omega + 10, & \omega > 4 \text{ rad/s} \end{cases}$$
Find the current flowing into the positively referenced input terminal.

****

From the frequency response, we can determine the impedance at each frequency of the input voltage:

$$R_{dc} = 4 \,\Omega$$
$$Z_2 = (5 \times 2 + 4)\underline{/50°} = 14\underline{/50°} \,\Omega$$
$$Z_5 = 1.5(5)^2 \underline{/(3 \times 5 + 10)°} = 37.5\underline{/25°} \,\Omega$$
$$Z_{10} = 1.5(10)^2 \underline{/(3 \times 10 + 10)°} = 150\underline{/40°} \,\Omega$$

Then, using $I = V/Z$ we can find the current phasor corresponding to each component of the input voltage, and from each current phasor the corresponding input current component.

$$I_{dc} = \frac{8}{4} = 2 \text{ A}$$
$$I_2 = \frac{-42\underline{/15°}}{14\underline{/50°}} = -3\underline{/-35°} \text{ A}$$

and so $i_2(t) = -3\cos(2t - 35°)$ A

$$I_5 = \frac{75\underline{/-130°}}{37.5\underline{/25°}} = 2\underline{/-155°} = -2\underline{/25°} \text{ A}$$

and so $i_5(t) = -2\cos(5t + 25°)$ A

and $$I_{10} = \frac{-120\underline{/50°}}{150\underline{/40°}} = -0.8\underline{/10°} \text{ A}$$
and so $i_{10}(t) = -0.8\cos(10t + 10°)$ A

Finally, $i(t) = I_{dc} + i_2(t) + i_5(t) + i_{10}(t)$
$= 2 - 3\cos(2t-35°) - 2\cos(5t+25°) - 0.8\cos(10t+10°)$ A

6-60

Determine the conditions on R relative to other circuit parameters such that the voltage magnitude across C is greater than the generator voltage magnitude. **at ω_o.** (SERIES RESONANCE)

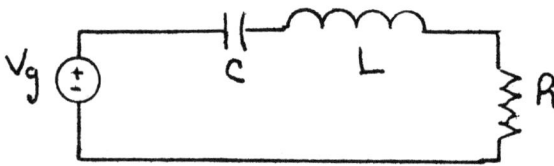

****

let $w = \omega_o = 1/\sqrt{LC}$

Using the voltage divider rule

$$\vec{V}_c = [-j/(wC)/(R+jwL - j/(wC))] \vec{V}_g, \text{ where } w = 1/\sqrt{LC}$$

At Resonance the j terms in the denominator cancel out and we are left with:

$$\vec{V}_c = (-j/(wC)/R) \vec{V}_g$$

Therefore if $1/(wC)/R$ is greater than 1, the $|V_c|$ is greater than $|V_g|$

$\therefore \quad R < \dfrac{1}{w_0 C}$

7
COMPLEX FREQUENCY

POLES, ZEROS, AND NETWORK FUNCTIONS

━━━━━━━━━━━━━━━━━━━━━━━━━━━━ 7-1

For a certain transfer function H(s), the Bode magnitude plot
   (a) Increases at a rate of 20 dB/decade at low frequencies.
   (b) Levels off at 0.2 rad/s.
   (c) Starts increasing at a rate of 20 dB/decade at 5 rad/s.
   (d) Starts decreasing at a rate of -20 dB/decade at 40 rad/s and continues decreasing at this rate for all greater frequencies.

Also, at 0.1 rad/s, the Bode plot has a value of 6 dB. Find H(s), given that it has no poles or zeros in the right-hand half of the s plane and that it has a positive scale factor.

****

From the change of slope specifications, the corner frequencies are 0.2, 5, and 40 rad/s. So H(s) has zeros or poles at -0.2, -5, and -40 in the s plane, the corresponding factors of which are (s+0.2), (s+5), and (s+40). Also, H(s) must have a zero at s=0 to produce the initially increasing slope. This zero corresponds to a numerator s factor. The (s+5) factor must also be in the numerator because at

5 rad/s the slope starts increasing. The two other factors $(s+0.2)$ and $(s+40)$ must be in the denominator because the slope changes more negatively at 0.2 and 40 rad/s. Further, the $(s+40)$ factor must be squared because at 40 rad/s the change in slope is by -40 dB/decade. Consequently, the transfer function has the form

$$H(s) = \frac{Ks(s+5)}{(s+0.2)(s+40)^2}$$

The scale factor K can be determined from the known value of 6 dB at 0.1 rad/s. The 6 dB corresponds to a magnitude of $10^{6/20} = 2$. Since 0.1 rad/s is less than all the corner frequencies, the Bode magnitude approximation at this frequency is that the s terms are negligible in the $(s+T)$ factors. The result is that

$$|H(j0.1)| = 2 \cong \frac{K|j0.1|(5)}{0.2(40)^2}$$

which solves to $K = 1280$. So,

$$H(s) = \frac{1280\,s(s+5)}{(s+0.2)(s+40)^2}$$

7-2

Find the poles of voltage V2(s). Calculate the sinusoidal steady state complex power supplied at ($\omega = 2$) and find the expression for the driving point impedance of the source for all complex frequencies.

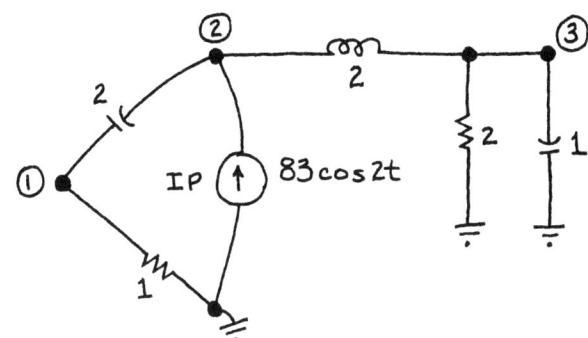

***

Write and solve the reference node model:

$$\begin{bmatrix} 1+2s & -2s & 0 \\ -2s & 2s+\tfrac{1}{2}s & -\tfrac{1}{2}s \\ 0 & -\tfrac{1}{2}s & \tfrac{1}{2}+s+\tfrac{1}{2}s \end{bmatrix} \begin{bmatrix} V1 \\ V2 \\ V3 \end{bmatrix} = \begin{bmatrix} 0 \\ IP \\ 0 \end{bmatrix} \quad \text{where } IP(s) = \left\{ \frac{83s}{s^2+4} \right\}$$

$$\frac{\begin{bmatrix} .25+1.5s+s^2+2s^3 & s+s^2+2s^3 & s \\ s+s^2+2s^3 & .5+1.5s+2s^2+2s^3 & .5+s \\ s & .5+s & .5+s+2s^2 \end{bmatrix} \begin{bmatrix} 0 \\ IP \\ 0 \end{bmatrix} = \begin{bmatrix} V1 \\ V2 \\ V3 \end{bmatrix}}{.25+2s+2s^2+2s^3}$$

$V2 = \left\{ \dfrac{83s}{s^2+4} \right\} \left\{ \dfrac{.5+1.5s+2s^2+2s^3}{.25+2s+2s^2+2s^3} \right\}$. The poles of V2 are: $\pm j2$, -0.1424, $-.4288 \pm j0.8331$.

$V2 \big|_{s=j\omega=j2} = 87.09 + j4.372 = 87.2 \underline{/2.87}$. The total power supplied is $(P+jQ) = \underbrace{3614}_{\text{watts}} + j\underbrace{181.4}_{\text{VARS}}$

$Zdp(s) = \left\{ \dfrac{.5+1.5s+2s^2+2s^3}{.25+2s+2s^2+2s^3} \right\}$ looking at entry (2,2) of the matrix inverse.

7-3

For the circuit below find $H(s) = V_2(s)/V_s$ and all its critical frequencies. Then find an expression for $|H(\omega)|$ and determine $|H(\omega)|_{max}$.

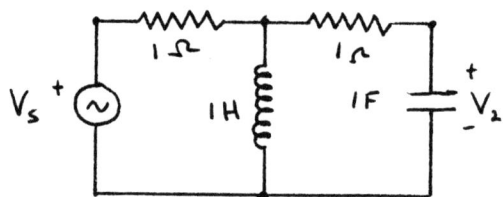

****

$$V_2(s) = \frac{I_2(s)}{s}$$

$$(s+1)I_1 - sI_2 = V_s$$
$$-sI_1 + (s+1+\tfrac{1}{s})I_2 = 0$$

$$-s^2 I_1 + (s^2+s+1)I_2 = 0$$

$$I_2 = \frac{\begin{vmatrix} s+1 & V_s \\ -s^2 & 0 \end{vmatrix}}{\begin{vmatrix} s+1 & -s \\ -s^2 & s^2+s+1 \end{vmatrix}} = \frac{s^2 V_s}{s^3 + 2s^2 + 2s + 1 - s^3}$$

$$I_2 = \frac{s^2 V_s}{2s^2 + 2s + 1}$$

$$V_2(s) = \frac{s V_s}{2s^2 + 2s + 1}$$

$$H(s) = \frac{s}{2s^2 + 2s + 1}$$

The critical frequencies are:

Zeros: $s = 0$, $s \to \infty$

Poles: $2s^2 + 2s + 1 = 0$

$$s = \frac{-2 \pm \sqrt{4-8}}{4} = \frac{-2 \pm j2}{4}$$

$$s = -0.5 \pm j0.5$$

$$H(\omega) = \frac{j\omega}{-2\omega^2 + j2\omega + 1} = \frac{j\omega}{1 - 2\omega^2 + j2\omega}$$

$$|H(\omega)| = \frac{\omega}{\sqrt{(1-2\omega^2)^2 + 4\omega^2}} = \frac{\omega}{\sqrt{1 + 4\omega^4}}$$

$|H(\omega)|_{max}$ occurs when $H(\omega)$ is real.

$H(\omega)$ is real when $1 - 2\omega^2 = 0$,

$$\omega = \frac{1}{\sqrt{2}}$$

$$|H(\omega)|_{max} = \frac{j(1/\sqrt{2})}{j2(1/\sqrt{2})} = \frac{1}{2} = H(1/\sqrt{2})$$

7-4

Determine the transimpedance $V_o(s)/I_1(s)$ for the network shown.

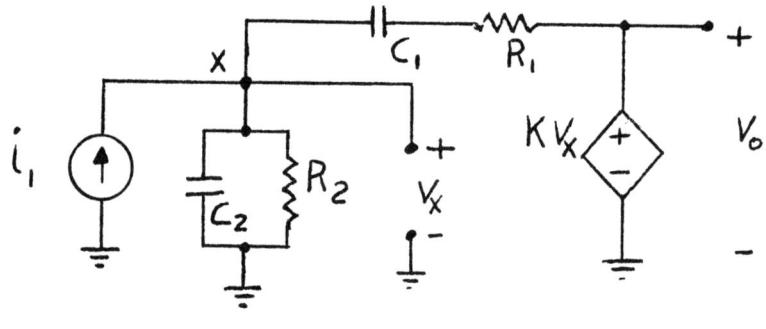

* *

NODE EQUATION AT X:

$$-I_1 + \frac{V_x(s)}{\frac{R_2}{sC_2 R_2 + 1}} + \frac{V_x(s) - KV_x(s)}{\frac{1}{sC_1} + R_1} = 0$$

$$V_x = I_1 \frac{R_2(sC_1 R_1 + 1)}{s^2 C_1 C_2 R_1 R_2 + s(C_1 R_1 + C_2 R_2 + (1-K)C_1 R_2) + 1}$$

$$V_o(s) = K V_x(s)$$

$$\frac{V_o}{I_1} = \frac{K}{C_2} \left[\frac{s + \frac{1}{R_1 C_1}}{s^2 + s \frac{C_1 R_1 + C_2 R_2 + (1-K)C_1 R_2}{C_1 C_2 R_1 R_2} + \frac{1}{C_1 C_2 R_1 R_2}} \right]$$

7-5

Find the poles and zeroes of the network function given by:

$$\frac{s^2 - 5s - 6}{s^3 + 6s^2 + 18s + 40}$$

****

Zeroes are roots of the numerator. We might ask if we can think of factors of -6 which combine as -5 and immediately, we say $1, -6$. So the numerator factors as $(s+1)(s-6)$ and the zeroes are $\boxed{-1, 6}$.

Poles are roots of the denominator. Because the order is odd, we know there is at least one real root. We might find it by synthetic division if we remember the technique. Alternatively, we guess at a factor and divide it until we find the one which divides without remainder.

$$s+2 \overline{\smash{\big)}\, \begin{array}{l} s^2 + 4s + 10 + \frac{20}{s+2} \\ s^3 + 6s^2 + 18s + 40 \\ \underline{s^3 + 2s^2} \\ 4s^2 + 18s \\ \underline{4s^2 + 8s} \\ 10s + 40 \\ \underline{10s + 20} \end{array}}$$

Remainder is obviously 40 if we try $s+0$, so guess we are half-way to root.

$$s+4 \overline{\smash{\big)}\, \begin{array}{l} s^2 + 2s + 10 \\ s^3 + 6s^2 + 18s + 40 \\ \underline{s^3 + 4s^2} \\ 2s^2 + 18s \\ \underline{2s^3 + 8s} \\ 10s + 40 \\ \underline{10s + 40} \end{array}}$$

We found it! Applying the quadratic formula to the quotient, we get

$$s = \frac{-2 \pm \sqrt{(2)^2 - 40}}{2}$$
$$= -1 \pm \sqrt{-36} = -1 \pm j6$$

So the poles are $\boxed{-4, -1 \pm j6}$

NATURAL AND FORCED RESPONSES DERIVED FROM THE NETWORK FUNCTION

7-6

The pole-zero diagram for the impedance of a circuit is shown below. If $z(0) = 12\Omega$, find $z(s)$. And if $v(t) = 24e^{-t}$, find $i(t)$.

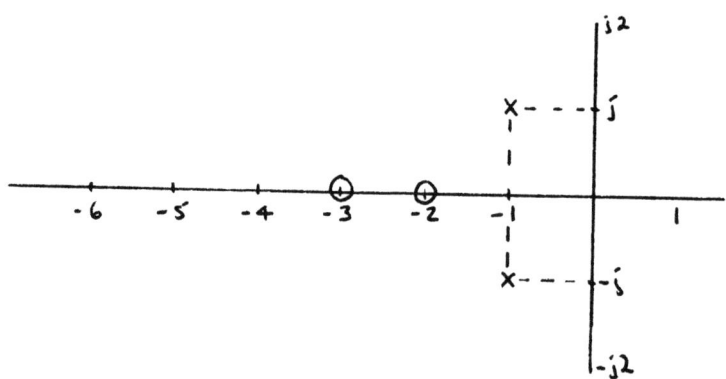

***

$$Z(s) = \frac{k(s+3)(s+2)}{(s+1-j)(s+1+j)} = \frac{k(s+3)(s+2)}{s^2+s+js+s+1+j-js-j+1}$$

$$= \frac{k(s+3)(s+2)}{s^2+2s+2}$$

$$Z(0) = 12 = \frac{k(3)(2)}{2} \qquad 3k = 12, \quad k = 4$$

$$Z(s) = \frac{4(s+3)(s+2)}{s^2+2s+2}$$

$$I = \frac{V}{Z} \qquad s = \sigma = -1 \qquad \text{for } v(t) = 24e^{-t}$$

$$I = \left.\frac{24(s^2+2s+2)}{4(s+3)(s+2)}\right|_{s=-1} = \frac{6(1-2+2)}{(2)(1)} = 3$$

$$i(t) = 3e^{-t} \quad A$$

7-7

For the network given, find for what value of α is the forced response zero, i.e., the effect of input does not appear in the output v_2?

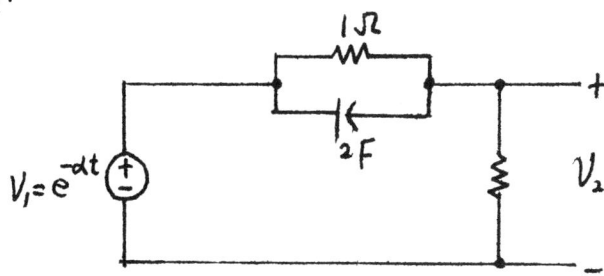

***

First obtain the voltage transfer function as follows.

$$\frac{V_2(s)}{V_1(s)} = \frac{1}{1 + \frac{1}{1+2s}} = \frac{1+2s}{2+2s}$$

Now $V_1(s) = \frac{1}{s+\alpha}$. In order that the forced response be zero, the denominator of $V_2(s)$ should not contain $s+\alpha$ factor.

$$V_2(s) = \frac{V_2(s)}{V_1(s)} \cdot V_1(s) = \frac{1+2s}{2(1+s)} \cdot \frac{1}{s+\alpha}$$

If we let $\alpha = \frac{1}{2}$, then $V_2(s) = \frac{1}{s+1}$

and the $s+\alpha$ term does not appear in $V_2(s)$.

7-8

The network shown represents an oscilloscope probe connected to an oscilloscope. The components C_2 and R_2 represent the input circuitry of the oscilloscope and C_1 and R_1 represent the probe.

a) Find the transfer function $\frac{V_o}{V_1}(s)$.

b) Find a relationship among the components that makes the natural response equal to zero for all time.

c) Suppose the excitation, $v_1(t)$, is a unit step of voltage. Sketch $v_o(t)$ if

1) $\dfrac{C_1}{C_1+C_2} = \dfrac{R_2}{R_1+R_2}$ 2) $\dfrac{C_1}{C_1+C_2} > \dfrac{R_2}{R_1+R_2}$ 3) $\dfrac{C_1}{C_1+C_2} < \dfrac{R_2}{R_1+R_2}$

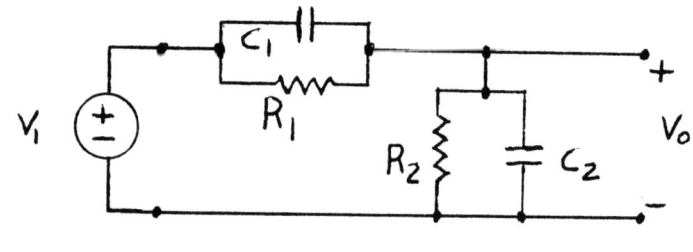

****

d) USE VOLTAGE DIVIDER:
$$\frac{V_o}{V_1}(s) = \frac{\frac{R_2}{sC_2 R_2 + 1}}{\frac{R_1}{sC_1 R_1 + 1} + \frac{R_2}{sC_2 R_2 + 1}}$$

$$\frac{V_o}{V_1} = \frac{C_1}{C_1+C_2}\left[\frac{s + \frac{1}{C_1 R_1}}{s + \frac{R_1+R_2}{R_1 R_2 (C_1+C_2)}}\right]$$

b) TO MAKE NATURAL RESPONSE ZERO, ELIMINATE THE POLE IN $\frac{V_o}{V_1}$ BY CAUSING IT TO CANCEL WITH THE ZERO.

$$-\frac{1}{C_1 R_1} = -\frac{R_1 + R_2}{R_1 R_2 (C_1 + C_2)}$$

THUS $\dfrac{C_1+C_2}{C_1} = \dfrac{R_1+R_2}{R_2}$ OR $1 + \dfrac{C_2}{C_1} = 1 + \dfrac{R_1}{R_2}$ OR $\dfrac{C_2}{C_1} = \dfrac{R_1}{R_2}$.

c) IF $V_1(t) =$ UNIT STEP, $V_1(s) = \dfrac{1}{s}$

$$V_o(s) = \frac{C_1}{C_1+C_2} \cdot \frac{s + \frac{1}{R_1 C_1}}{s\left[s + \frac{R_1+R_2}{R_1 R_2 (C_1+C_2)}\right]} = \frac{K_1}{s} + \frac{K_2}{s + \frac{R_1+R_2}{R_1 R_2 (C_1+C_2)}}$$

$$K_1 = \frac{R_2}{R_1+R_2} \quad \text{AND} \quad K_2 = \frac{C_1}{C_1+C_2} - \frac{R_2}{R_1+R_2}$$

$$V_o(t) = \frac{R_2}{R_1+R_2} + \left[\frac{C_1}{C_1+C_2} - \frac{R_2}{R_1+R_2}\right]\varepsilon^{-\frac{t}{\tau}} \quad , \quad t>0$$

WHERE $\tau = \dfrac{R_1 R_2 (C_1+C_2)}{R_1+R_2}$

1) IF $\dfrac{C_1}{C_1+C_2} = \dfrac{R_2}{R_1+R_2}$ THEN $V_o(t) = \dfrac{R_2}{R_1+R_2} = \dfrac{C_1}{C_1+C_2}$

2) AND 3)

$$V_o(t=0^+) = \frac{C_1}{C_1+C_2}$$

$$V_o(t\to\infty) = \frac{R_2}{R_1+R_2}$$

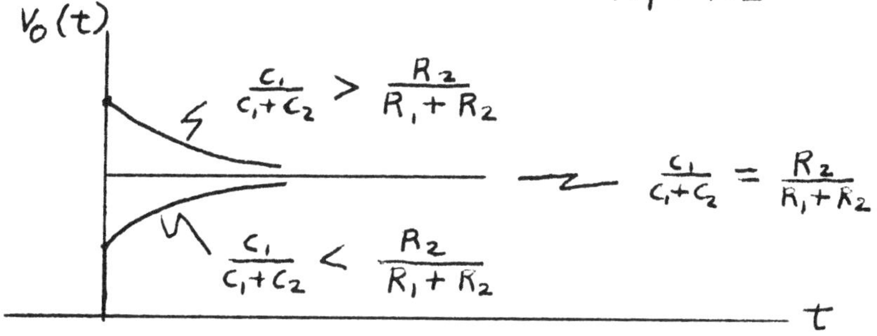

7-9

Find the form of the natural response of the circuit shown by using complex impedance or admittance functions at:
a) terminals a-b.
b) terminals c-d with the circuit cut at the point marked "x".

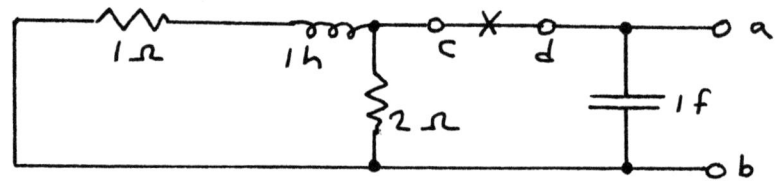

***

a)
$$Z_{ab} = \frac{\left[\frac{2/s}{2+1/s}\right][s+1]}{\frac{2/s}{2+1/s} + s + 1} = \frac{\frac{2}{2s+1}(1+s)}{\frac{2}{2s+1} + 1 + s} = \frac{s+1}{s^2 + 3/2\, s + 3/2}$$

want poles of Z_{ab} since ab normally open

$$S = \frac{-3/2 \pm \sqrt{9/4 - 6}}{2} = -3/4 \pm j\sqrt{15}/4$$

$$\boxed{\text{response} = A e^{-3/4\, t} \cos\frac{\sqrt{15}}{4} t + B e^{-3/4\, t} \sin\frac{\sqrt{15}}{4} t}$$

b)
$$Z_{cd} = \frac{1}{s} + \frac{2(1+s)}{2+(1+s)}$$

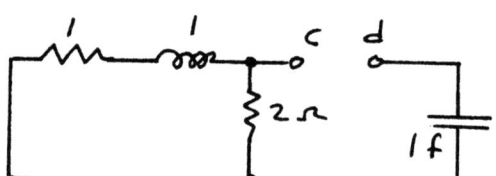

$$Z_{cd} = \frac{2s^2 + 3s + 3}{2(s+3)} = \frac{s^2 + 3/2\, s + 3/2}{s+3}$$

want zeros of Z_{cd} since cd normally shorted

$$S = -3/4 \pm j\frac{\sqrt{15}}{4} \quad \left(\text{as in a) above}\right)$$

response same as part a)

8
TWO PORT NETWORKS

DEFINITION OF A TWO-PORT; PORT CONDITIONS

8-1

(a) Find the voltage transfer function for the network given.

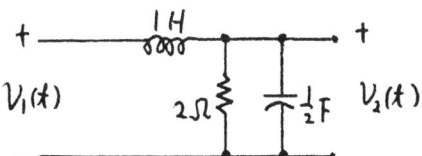

(b) Find the input impedance for the RC ladder network given.

****

(a) $$\frac{V_2(s)}{V_1(s)} = \frac{\frac{2 \cdot \frac{2}{s}}{2 + 2/s}}{s + \frac{2 \cdot \frac{2}{s}}{2 + \frac{2}{s}}} = \frac{2}{s^2 + s + 2}$$

(b) $$Z_{in}(s) = \frac{1}{s} + \frac{1 \cdot (1 + \frac{1}{2s})}{1 + 1 + \frac{1}{2s}}$$

$$= \frac{2s^2 + 5s + 1}{s(4s+1)}$$

TWO-PORT PARAMETERS

8-2

For the circuit given, calculate the y-parameter, y_{21}.

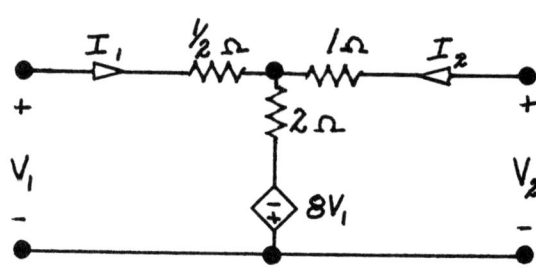

****

$y_{21} = \dfrac{I_2}{V_1}\bigg|_{V_2=0}$ ∴ USE THE FOLLOWING CIRCUIT.

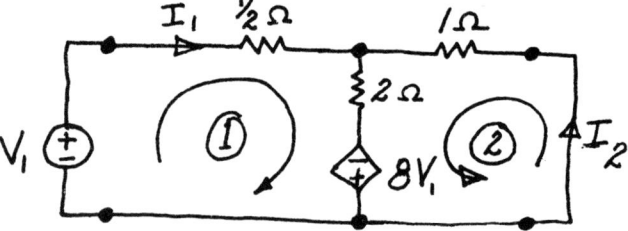

KVL①:

$\tfrac{1}{2} I_1 + 2(I_1 + I_2) = 8V_1 + V_1$

$\Rightarrow 5I_1 + 4I_2 = 18 V_1$

KVL②:

$1(I_2) + 2(I_1 + I_2) = 8V_1$

$\Rightarrow 2I_1 + 3I_2 = 8V_1$

Now SOLVE FOR I_2 IN TERMS OF V_1.

$\Rightarrow I_2 = \dfrac{4V_1}{7}$ OR $\dfrac{I_2}{V_1}\bigg|_{V_2=0} = \dfrac{4}{7}$ S.

8-3

Find the h-parameters for the circuit shown. (Note: $2I_a$ is a current-controlled current source.)

$V_1 = h_{11} I_1 + h_{12} V_2$

$I_2 = h_{21} I_1 + h_{22} V_2$

****

First, set $I_1 = 0$
apply V_2

$I_1 = I_a = 0 \rightarrow I_2 = \dfrac{V_2}{4+6} = .1 V_2 \rightarrow \boxed{h_{22} = 0.15 \text{ (}\mho\text{)}}$

$V_1 = \dfrac{6}{4+6} V_2 = .6 V_2 \rightarrow \boxed{h_{12} = 0.6} \text{ unitless}$

Second, set $V_2 = 0$
apply I_1

$I_1 = I_a$

$V_1 = 2 I_1 + (I_1 + 2 I_1) \dfrac{4 \times 6}{4+6} = 9.2 I_1 \rightarrow \boxed{h_{11} = 9.2 \, \Omega}$

$I_2 = 2 I_1 - \dfrac{(I_1 + 2 I_1) \frac{4 \times 6}{4+6}}{4} = 0.2 I_1 \rightarrow \boxed{h_{21} = 0.2} \text{ unitless}$

$\begin{bmatrix} V_1 \\ I_2 \end{bmatrix} = \begin{bmatrix} 9.2 & 0.6 \\ 0.2 & 0.1 \end{bmatrix} \begin{bmatrix} I_1 \\ V_2 \end{bmatrix}$

8-4

Find the h-parameters for the two-port shown below.

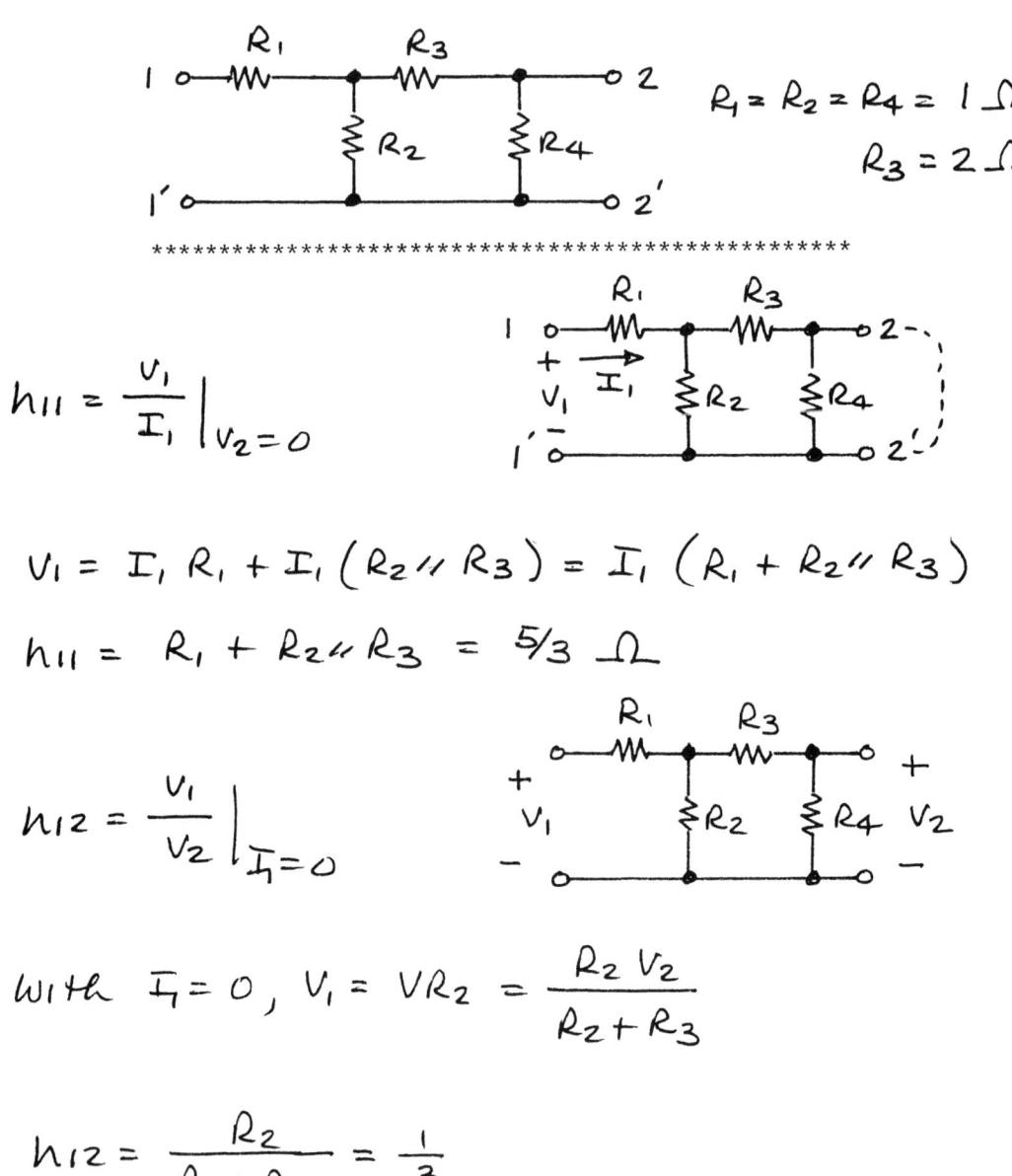

$R_1 = R_2 = R_4 = 1\,\Omega$
$R_3 = 2\,\Omega$

$h_{11} = \dfrac{V_1}{I_1}\bigg|_{V_2=0}$

$V_1 = I_1 R_1 + I_1 (R_2 // R_3) = I_1 (R_1 + R_2 // R_3)$

$h_{11} = R_1 + R_2 // R_3 = 5/3\,\Omega$

$h_{12} = \dfrac{V_1}{V_2}\bigg|_{I_1=0}$

With $I_1 = 0$, $V_1 = V_{R_2} = \dfrac{R_2 V_2}{R_2 + R_3}$

$h_{12} = \dfrac{R_2}{R_2 + R_3} = \dfrac{1}{3}$

$$h_{21} = \left.\frac{I_2}{I_1}\right|_{V_2=0}$$

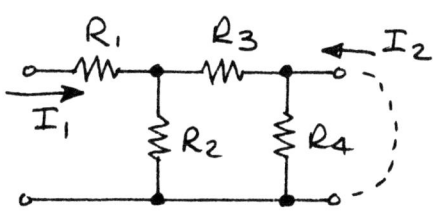

With $V_2=0$, $I_2 = I_{R_3}$ since R_4 is shorted.

$$I_{R_3} = \frac{-R_2 I_1}{R_2 + R_3} \qquad h_{21} = \frac{-R_2}{R_2+R_3} = -\frac{1}{3}$$

$$h_{22} = \left.\frac{I_2}{V_2}\right|_{I_1=0}$$

With $I_1=0$, no current flows in R_1,
KCL at the top of R_4

$$I_2 = \frac{V_2}{R_4} + \frac{V_2}{R_2+R_3} = V_2 \frac{R_2+R_3+R_4}{R_4(R_2+R_3)}$$

$$h_{22} = \frac{R_2+R_3+R_4}{R_4(R_2+R_3)} = \frac{4}{3} \text{ mho}$$

$$\boxed{\begin{array}{ll} h_{11} = \frac{5}{3}\,\Omega & h_{21} = -\frac{1}{3} \\ h_{12} = \frac{1}{3} & h_{22} = \frac{4}{3}\,\mho \end{array}}$$

8-5

The two port network below contains an ideal transformer.

a.) Find the Z parameters of the network.
b.) Is the circuit bi-lateral?
c.) Draw a "Y" equivalent circuit of the network using only resistors if possible.

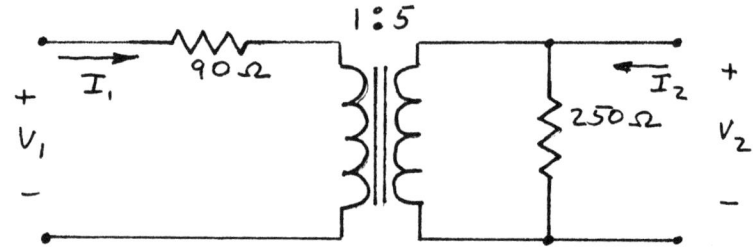

****

a) TWO-PORT EQNS:
$$V_1 = Z_{11} I_1 + Z_{12} I_2$$
$$V_2 = Z_{21} I_1 + Z_{22} I_2$$

$$Z_{11} = \left.\frac{V_1}{I_1}\right|_{I_2=0} = 90 + \frac{250}{N^2} \quad ; \quad N = \text{TURNS RATIO}$$
$$N = 5$$

$$= 90 + \frac{250}{25} = 100 \, \Omega \quad \text{ANS}$$

$$Z_{12} = \left.\frac{V_1}{I_2}\right|_{I_1=0}$$

$$V_1 = \frac{V_2}{N} = \frac{250 I_2}{5} = 50 I_2$$

$$\therefore Z_{12} = \frac{V_1}{I_2} = 50 \, \Omega \quad \text{ANS}$$

$$Z_{21} = \left.\frac{V_2}{I_1}\right|_{I_2=0}$$

$$V_2 = 250 \frac{I_1}{5} = 50 I_1$$

$$\therefore Z_{12} = \frac{V_2}{I_1} = 50 \, \Omega \quad \text{ANS}$$

$$Z_{22} = \frac{V_2}{I_2}\bigg|_{I_1=0} = 250\,\Omega \quad\quad \text{ANS}$$

b) SINCE $Z_{12} = Z_{21}$, CIRCUIT IS BILATERAL

c) SINCE CIRCUIT IS BILATERAL IT CAN BE REALIZED WITH PASSIVE ELEMENTS THUS

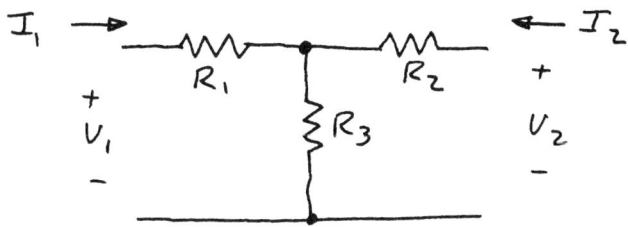

USING SAME DEFINITIONS

$$Z_{11} = R_1 + R_3 = 100$$

$$Z_{12} = \frac{V_1}{I_2}\bigg|_{I_1=0} = R_3 = 50\,\Omega \quad \text{ANS}$$

$$Z_{22} = R_2 + R_3 = 250$$

$$\therefore R_1 = 100 - R_3 = 100 - 50 = 50\,\Omega \quad \text{ANS}$$

AND
$$R_2 = 250 - R_3 = 250 - 50 = 200\,\Omega \quad \text{ANS}$$

8-6

The impedance parameters for the circuit below are: $z_{11} = 6\,\Omega$, $z_{12} = j4\Omega$, $z_{21} = j4\Omega$, $z_{22} = j2\Omega$, at $\omega = 10$ rad/s. Find the Thevenin equivalent circuit of the Z parameter network plus source and then find $i(t)$ if the source voltage is now changed to $v_s(t) = 20 \cos 10t$.

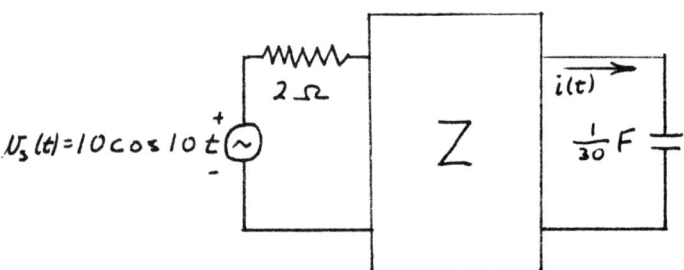

****

The open circuit voltage is given by

$$V_{oc} = V_{th} = \frac{z_{21}}{z_{11} + z_s} V_s$$

$V_s = 10 \qquad z_s = 2$

$$V_{th} = \frac{j4}{8} \cdot 10 = j5$$

The Thevenin impedance is:

$$Z_{th} = z_{22} - \frac{z_{12} z_{21}}{z_{11} + z_s} = j2 - \frac{(j4)(j4)}{8}$$

$$Z_{th} = 2 + j2$$

In the frequency domain this becomes

[circuit: $j5$ source, $2+j2$ impedance, $-j3$ capacitor]

and in the time domain:

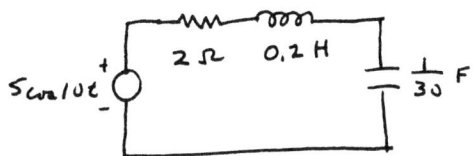

When v_s becomes $20\cos 10t$, the only change was in magnitude of the source so the Thevenin equivalent, in the frequency domain becomes:

$$I = \frac{j10}{2+j2-j3} = \frac{j10}{2-j} = \frac{10\angle 90°}{\sqrt{4+1}\angle \tan^{-1}\frac{-1}{2}}$$

$$I = \frac{10}{\sqrt{5}}\angle 90°-(-26.57°) = 4.47\angle 116.6° \text{ A}$$

$$i(t) = 4.47 \cos(10t + 116.6°) \text{ A}$$

8-7

Find all h parameters for this network.

[Circuit diagram: Input port with I_1 entering through a 2Ω resistor, V_1 across input. A 3Ω resistor and dependent voltage source $6V_2$ in series form the shunt branch. Output port has I_2 entering, V_2 across output, with 1Ω resistor in the bottom output lead.]

$$h_{11} = \left.\frac{V_1}{I_1}\right|_{V_2=0} = 2 + \frac{1(3)}{1+3} = 2 + \frac{3}{4} = \frac{11}{4}\,\Omega$$

$$h_{12} = \left.\frac{V_1}{V_2}\right|_{I_1=0}, \quad \text{KVL:} \quad V_1 = 3I_2 + 6V_2$$
$$V_2 = 4I_2 + 6V_2 \Rightarrow I_2 = -\frac{5}{4}V_2$$

$$= \frac{3\left(-\frac{5}{4}V_2\right) + 6V_2}{V_2} = \frac{9}{4}$$

$$h_{21} = \left.\frac{I_2}{I_1}\right|_{V_2=0} = \frac{-\frac{3}{4}I_1}{I_1} = -\frac{3}{4}$$

$$h_{22} = \left.\frac{I_2}{V_2}\right|_{I_1=0} = \frac{-\frac{5}{4}V_2}{V_2} = -\frac{5}{4}\,\mho$$

8-8

Find the z-parameter for the resistor network containing a controlled source as shown in the following figure.

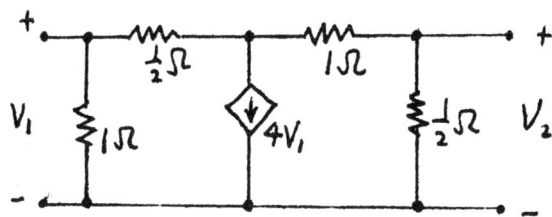

****

To find z_{11}, z_{21}, use the following network

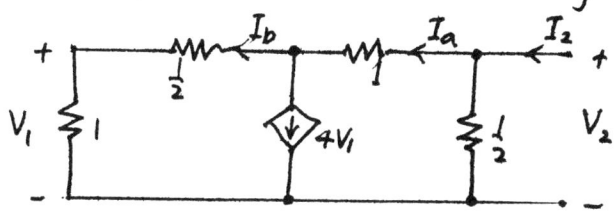

$I_1 = \frac{V_1}{1} + I_a$, $I_a = I_b + 4V_1$, $V_1 = \frac{1}{2} I_a + I_b(\frac{3}{2})$

Eliminate I_a, I_b, $\quad \frac{V_1}{I_1} = \frac{2}{9} = z_{11}$

Also $z_{21} = \frac{V_2}{I_1}\Big|_{I_2=0} = -\frac{1}{18} \quad$ as $V_2 = I_b(\frac{1}{2})$

To find z_{22}, z_{12}, use the following network.

$I_2 = I_a + \frac{V_2}{\frac{1}{2}} = I_a + 2V_2$, $V_2 = I_a \cdot 1 + I_b(\frac{3}{2})$

$I_a = I_b + 4V_1 = 5I_b$, $I_b = I_a - 4V_1$, $V_1 = I_b \cdot 1$

$z_{22} = \frac{V_2}{I_2}\Big|_{I_1=0} = \frac{13}{36}$, $z_{12} = \frac{V_1}{I_2}\Big|_{I_1=0} = \frac{1}{7}$

TRANSFORMERS

8-9

For the transformer system shown, determine
a. the turns ratio
b. the value of Rab
c. the current supplied by the source

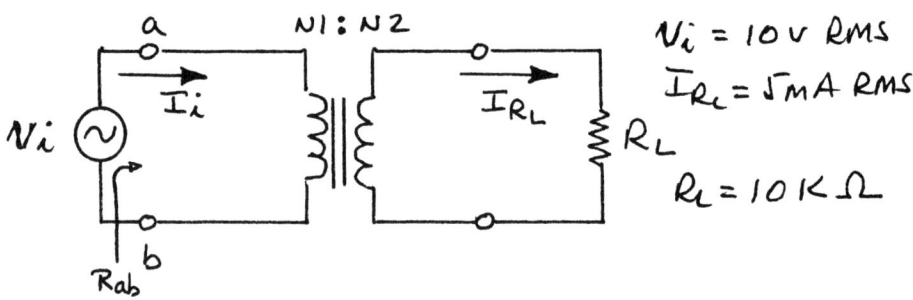

****

a. $V_{R_L} = I_{R_L} R_L = (5mA)(10K) = 50 V$ Rms

$\dfrac{N_1}{N_2} = \dfrac{V_i}{V_{R_L}} = \dfrac{10}{50} = \dfrac{1}{5}$ $\boxed{N_1 : N_2 = 1 : 5}$

b. $R_{ab} = \left(\dfrac{N_1}{N_2}\right)^2 R_L = \left(\dfrac{1}{5}\right)^2 10K = 400 \Omega$

$\boxed{R_{ab} = 400 \Omega}$

c. $I_i = \left(\dfrac{N_2}{N_1}\right) I_{R_L} = (5)(5mA) = 25 mA$

$\boxed{I_i = 25 mA}$

8-10

Find the turns ratio, n, necessary to provide maximum power to the 4Ω resistor.

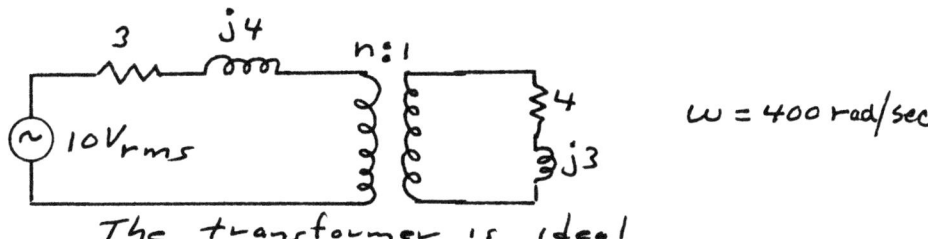

$\omega = 400$ rad/sec

The transformer is ideal

****

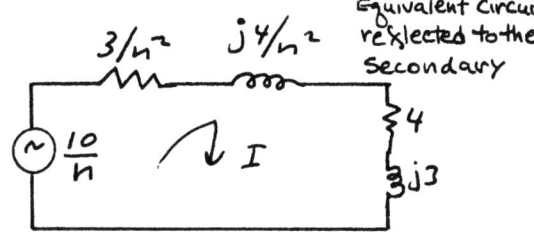

Equivalent circuit reflected to the secondary

Direct method

$$I = \frac{10/n}{(3/n^2 + 4) + j(4/n^2 + 3)}$$

$$P_4 = \frac{4 \times 10^2/n^2}{\left(\frac{3}{n^2}+4\right)^2 + \left(\frac{4}{n^2}+3\right)^2} = \frac{400 n^2}{25 + 48n^2 + 25n^4}$$

$$\frac{dP_4}{dn} = 0 = 400\left[\frac{2n(25+48n^2+25n^4) - n^2(96n + 100n^3)}{25+48n^2+25n^4}\right]$$

$$-50n^5 + 50n = 0$$
$$n^4 = 1 \rightarrow \boxed{n = 1}$$

By maximum power transfer theorem

$$n^2 = |Z_s|/|Z_L| = 5/5 = 1$$

$$\boxed{n = 1}$$

Note: This is maximum power transfer when only the turns ratio is adjustable.

8-11

Find the input impedance for the network shown. Simplify the expression found for the case in which the coefficient of coupling equals 1 (k = 1) and $L_1 = L_2$.

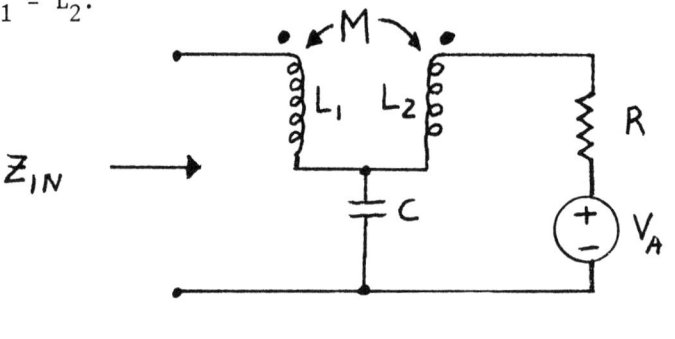

* *

REMOVE THE INDEPENDENT SOURCE V_A, EXCITE THE NETWORK, AND WRITE LOOP EQUATIONS:

$$Z_{IN} = \frac{V(s)}{I_1(s)}$$

$$V(s) = I_1\left(sL_1 + \frac{1}{sC}\right) - I_2 \frac{1}{sC} - I_2 sM$$

$$0 = -I_1 \frac{1}{sC} - I_1 sM + I_2\left(sL_2 + R + \frac{1}{sC}\right)$$

$$I_1(s) = \frac{\begin{vmatrix} V & -\frac{1}{sC} - sM \\ 0 & sL_2 + R + \frac{1}{sC} \end{vmatrix}}{\begin{vmatrix} (sL_1 + \frac{1}{sC}) & -(\frac{1}{sC} + sM) \\ -(\frac{1}{sC} + sM) & (sL_2 + R + \frac{1}{sC}) \end{vmatrix}}$$

$$I_1(s) = \frac{V(s)\left(sL_2 + R + \frac{1}{sC}\right)}{\left(sL_1 + \frac{1}{sC}\right)\left(sL_2 + R + \frac{1}{sC}\right) - \left(\frac{1}{sC} + sM\right)^2}$$

$$\frac{V(s)}{I_1(s)} = \frac{s^3 C(L_1 L_2 - M^2) + s^2 C L_1 R + s(L_1 + L_2 - 2M) + R}{s^2 C L_2 + sCR + 1} = Z_{IN}$$

IF $k = 1 = \dfrac{M}{\sqrt{L_1 L_2}}$

THEN Z_{IN} SIMPLIFYS TO:

$$Z_{IN} = R \dfrac{s^2 + \frac{1}{LC}}{s^2 + s\frac{R}{L} + \frac{1}{LC}}$$

8-12

For the circuit below determine the output voltage $v_o(t)$ if the input voltage $v_i(t)$ is 150 sin 377t. The transformer is ideal.

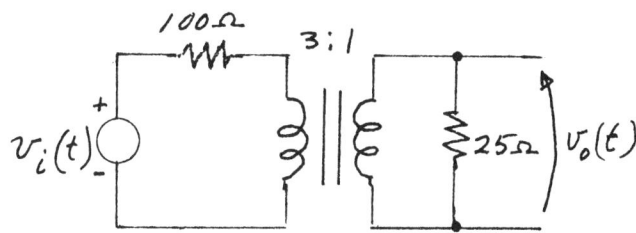

****

Reflected impedance at the primary is $n^2 R_L = 9(25)$
$= 225\,\Omega$

Thus the primary voltage is:

$$v_p(t) = \dfrac{225}{100 + 225} v_i(t) = 103.8 \sin 377t$$

and $v_o(t) = \frac{1}{3} v_p(t)$

$$\underline{v_o(t) = 34.6 \sin 377t}$$

8-13

For the coupled circuit shown, $R_1 = 5$ ohms, $L_{22} = 1$ H, $R_2 = 500$ ohms. The coils are wound on the same iron core and the number of turns in the secondary is ten times the number of turns in the primary. If the co-efficient of coupling, $k = \frac{1}{2}$, find the current I_1 in phasor form if $\overset{\circ}{V}_1 = 100 \underline{/0°}$ V, $\overset{\circ}{V}_2 = 1000 \underline{/0°}$ V and the frequency is 60 Hz.

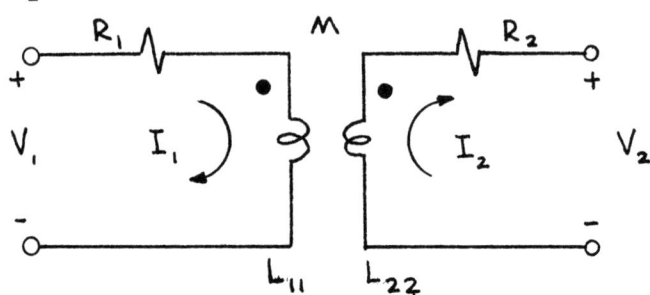

***

The assumed positive direction of I_2 being out of the dot terminal results in a negative M. The mesh equations on the primary and secondary sides are:

$$\dot{V}_1 = (R_1 + j\omega L_{11})\dot{I}_1 - j\omega M \dot{I}_2$$
$$-\dot{V}_2 = -j\omega M \dot{I}_1 + (R_2 + j\omega L_{22})\dot{I}_2$$

We determine L_{11} and M as follows:
Inductance $\propto$ (turns)^2 hence $L_{11} \propto N_1^2$, $L_{22} \propto N_2^2$ (will be same constant of proportionality)

$$\frac{L_{11}}{L_{22}} = \left(\frac{N_1}{N_2}\right)^2 = \left(\frac{1}{10}\right)^2; \quad L_{11} = 0.01 \text{ H}.$$

Since co-efficient of coupling, $k = \frac{M}{\sqrt{L_{11} L_{22}}}$,

$$M = k \cdot \sqrt{L_{11} L_{22}} = \frac{1}{2}\sqrt{1 \times 0.01} = 0.05 \text{ H}$$

Now substituting in the above eqns. we have,

$$100\underline{/0°} = (5 + j\,3.77)\dot{I}_1 - j\,18.85\,\dot{I}_2$$
$$-1000\underline{/0°} = -j\,18.85\,\dot{I}_1 + (500 + j\,377)\dot{I}_2$$

resulting in the polar form of the equation,

$$100\underline{/0°} = 6.26\underline{/37°}\,\dot{I}_1 + 18.85\underline{/-90°}\,\dot{I}_2$$
$$-1000\underline{/0°} = 18.85\underline{/-90°}\,\dot{I}_1 + 626.2\underline{/37°}\,\dot{I}_2$$

Using Cramer's rule:

$$\dot{I}_1 = \frac{\begin{vmatrix} 100\underline{/0°} & 18.85\underline{/-90°} \\ -1000\underline{/0°} & 626.2\underline{/37°} \end{vmatrix}}{\begin{vmatrix} 6.26\underline{/37°} & 18.85\underline{/-90°} \\ 18.85\underline{/-90°} & 626.2\underline{/37°} \end{vmatrix}} = \frac{62620\underline{/37°} + 18850\underline{/-90°}}{3920\underline{/74°} + 335\underline{/0°}}$$

$$= \frac{50000 + j\,18850}{1436 + j\,3769} = \frac{53435\underline{/20.7°}}{4033\underline{/69.2°}} = 13.25\underline{/-48.5°}\,A$$

Similarly, $\dot{I}_2 = 1.325\underline{/-228.5°}$ A.

8-14

An amplifier with an output impedance of 10,000 ohms is to be transformer coupled to an 8 ohm speaker. What turns ratio is required to achieve maximum power transfer to the speaker?

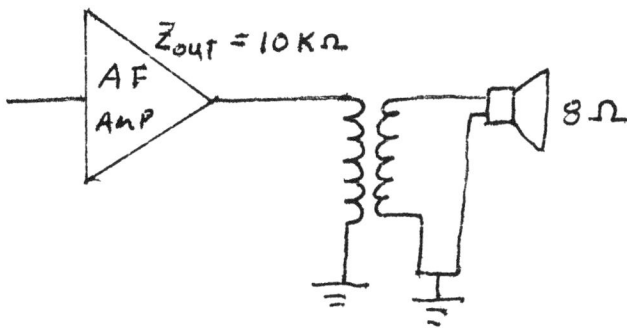

** * ** * ** *** ** * ** *** ** ** *** ** ***** ***** **

$$Z_P = a^2 Z_S$$

$$a = \sqrt{\frac{Z_P}{Z_S}} = \sqrt{\frac{10,000}{8}} = 35.36$$

Turns ratio should be ≈ 35:1

8-15

For the transformer-connected load shown:

A. Find the input admittance Y_{in} at 50 Hz.

B. If we say that the transformer approximates an ideal transformer at any frequency for which the reflected impedance is less than 1/10 the impedance of the magnetizing inductance, what is the lowest frequency meeting this criterion? (Such considerations determine the low frequency effectiveness of an audio amplifier using output transformers.)

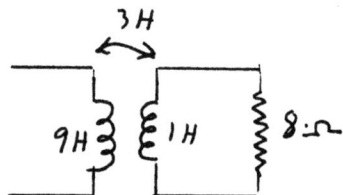

***

We have perfect coupling between the windings, since $M = \sqrt{L_1 L_2}$. Thus, we may expect input admittance to be magnetizing inductance in parallel with the 8 ohms converted by an ideal transformer of 3:1 turns ratio, since $L_1 = K n_1^2$. If we don't remember that, we can write equations, assuming a voltage V_1 connected to the primary

$$V_1 = j\omega 9 I_1 - j\omega 3 I_2$$
$$0 = -j\omega 3 I_1 + (8+j\omega) I_2 \quad \text{Solving for } I_1$$

$$I_1 = \frac{\begin{vmatrix} V_1 & -j\omega 3 \\ 0 & 8+j\omega \end{vmatrix}}{\begin{vmatrix} j\omega 9 & -j\omega 3 \\ -j\omega 3 & 8+j\omega \end{vmatrix}} = \frac{V_1 (8+j\omega)}{j\omega 72 - 9\omega^2 + 9\omega^2} = \left(\frac{-j}{9\omega} + \frac{1}{72}\right) V_1$$

$Y_{in} = I_1/V_1 = \boxed{\dfrac{-j}{9\omega} + \dfrac{1}{72}}$. Indeed, 8Ω was multiplied by $(3)^2$ and is shunted by $j\omega L_1$.

B. We require $\dfrac{1}{9\omega} \times 0.1 = \dfrac{1}{72}$ $\quad \omega = \dfrac{7.2}{9} \quad f = \dfrac{\omega}{2\pi} = \dfrac{7.2}{18\pi}$

$$= \boxed{.127 \text{ Hz}}$$

8-16

Find **V** in the shown circuit.

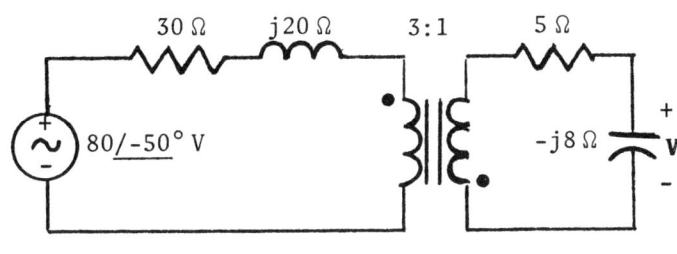

*****

Since the turns ratio is specified instead of the winding inductances, the transformer should be assumed to be ideal. A good start in the analysis is to use reflected impedance and voltage division to obtain the voltage across the primary winding. The reflected impedance, which replaces the primary winding, is the square of the inverse of the turns ratio times the total impedance of the secondary circuit:

$$3^2(5-j8) = 45 - j72 \; \Omega$$

Then by voltage division, the voltage across this impedance is

$$\frac{45-j72}{30+j20+45-j72} \times 80\underline{/-50°} = \frac{84.9\underline{/-58.0°}}{91.3\underline{/-34.7°}} \times 80\underline{/-50°}$$
$$= 74.4 \underline{/-73.3°} \; V$$

which is also the primary winding voltage, referenced positive at the dot. From the turns ratio, the secondary winding voltage referenced positive at the dot, is

$$\frac{74.4\underline{/-73.3°}}{3} = 24.8\underline{/-73.3°} \; V$$

Finally, by voltage division, the desired voltage is

$$V = -\frac{-j8}{5-j8}(24.8\underline{/-73.3°}) = \frac{8\underline{/90°}}{9.43\underline{/-58.0°}} \times 24.8\underline{/-73.3°} = 21.0\underline{/74.7°} \; V$$

8-17

Find I_1 and I_2.

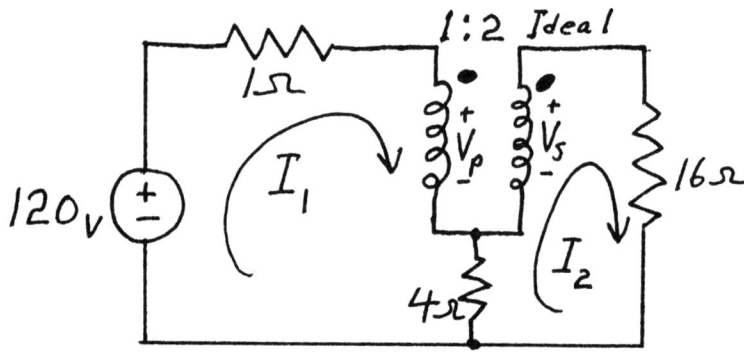

Using mesh analysis we write two mesh equations.

$$5I_1 + V_p - 4I_2 = 120$$

$$-4I_1 - V_s + 20I_2 = 0$$

Now use transformer relations $V_s = 2V_p$ and $I_1 = 2I_2$. Substitute into equations above.

$$5(2I_2) + V_p - 4I_2 = 120 \implies 6I_2 + V_p = 120$$

$$-4(2I_2) - 2V_p + 20I_2 = 0 \implies 12I_2 - 2V_p = 0$$

Solving these we obtain $I_2 = 10$ A and $I_1 = 20$ A

8-18

For the mutually coupled windings given, use the right-hand-rule and Lenz's law to assign the dot convention symbols.

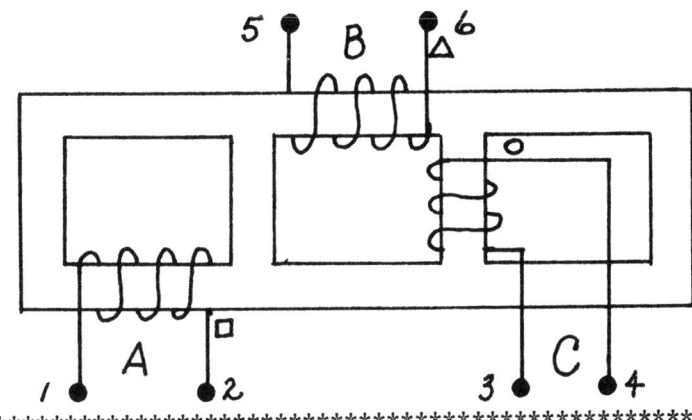

CURRENT INTO A2 PRODUCES FLUX IN C THAT MUST RESULT IN A CURRENT LEAVING C3 TO OPPOSE CAUSING FLUX.

CURRENT INTO B6 PRODUCES FLUX IN A THAT MUST RESULT IN A CURRENT LEAVING A1 TO OPPOSE CAUSING FLUX.

CURRENT INTO C4 PRODUCES FLUX IN B THAT MUST RESULT IN A CURRENT LEAVING B6 TO OPPOSE CAUSING FLUX.

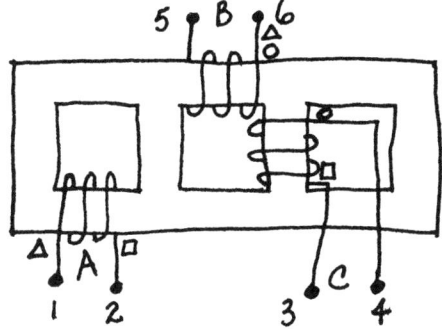

8-19

a) Find n for maximum power in the 2 ohm resistor.
b) Find the power in the 2 ohm resistor if n=4.

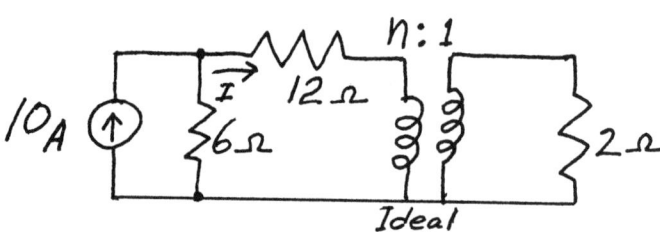

****

a) Find the Thevenin Equivalent of the circuit to the left of the transformer. It is

To obtain the maximum power from this circuit the load across it should be 18Ω.

The Impedance seen looking into the n turn side of the transformer is $n^2(2)$

$$n^2(2) = 18$$
$$n = 3$$

for $n=3$
Power in $2\Omega = 50w$

b) If $n=4$ the impedance seen looking into the transformer is $4^2(2) = 32\Omega$

The power in 2Ω is same as power into transformer.

Using current division $I = 10 \frac{6}{6+44} = 6/5 \, A$

$$P = I^2(32) = \underline{46.08 \, w}$$

or calculating Power in secondary $P = (6/5 \times 4)^2 (2) = 46.08_w$

8-20

The current flowing in the 200 Ω load in the circuit below is 165° out of phase with the 20 volt source. Find the coupling coefficient k and the current i(t).

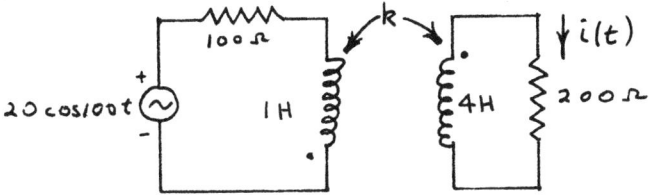

****

$$M = k\sqrt{L_1 L_2} = k\sqrt{1\cdot 4} = 2k$$

$$(100 + j100)\, I_1 + j200k\, I_2 = 20$$
$$j200k\, I_1 + (200 + j400)\, I_2 = 0$$

$$(1+j)\, I_1 + j2k\, I_2 = 0.2$$
$$jk\, I_1 + (1+j2)\, I_2 = 0$$

$$I_2 = \frac{\begin{vmatrix} 1+j & 0.2 \\ jk & 0 \end{vmatrix}}{\begin{vmatrix} 1+j & j2k \\ jk & 1+j2 \end{vmatrix}} = \frac{-j0.2k}{1+j+j2-2+2k^2}$$

$$= \frac{j0.2k}{1-2k^2 - j3} = \frac{0.2k\,\angle 90°}{\sqrt{(1-2k^2)^2 + 9}\,\angle\tan^{-1}\left(\frac{-3}{1-2k^2}\right)}$$

$$90° - \tan^{-1}\left(\frac{-3}{1-2k^2}\right) = 165°$$

$$\tan^{-1}\left(\frac{-3}{1-2k^2}\right) = -75°$$

$$\frac{-3}{1-2k^2} = \tan(-75°) = -(2+\sqrt{3})$$

$$\frac{3}{1-2k^2} = 2+\sqrt{3} \qquad 1-2k^2 = \frac{3}{2+\sqrt{3}}$$

$$2k^2 = 1 - \frac{3}{2+\sqrt{3}} \qquad k^2 = \frac{1}{2} - \frac{3/2}{2+\sqrt{3}}$$

$$k = \sqrt{\frac{1}{2} - \frac{3/2}{2+\sqrt{3}}} = 0.313$$

$$I = I_2 = \frac{(0.2)\cdot(0.313)\,\underline{/90°}}{\sqrt{[1-2\cdot(.313)]^2 + 9}\,\underline{/-75°}}$$

$$= 0.0202\,\underline{/165°}\ A$$

$$i(t) = 20.2\,\cos(100t + 165°)\ mA$$

9
STATE VARIABLE ANALYSIS

SOLUTION OF THE VECTOR-MATRIX STATE EQUATION

━━ 9-1

Determine the characteristic values and one set of characteristic vectors for the matrix:

$$A = \begin{bmatrix} 5 & 4 \\ 1 & 2 \end{bmatrix}$$

Hence obtain the matrix e^{At}.

****

The characteristic equation is

$$|A - \lambda I| = \begin{vmatrix} 5-\lambda & 4 \\ 1 & 2-\lambda \end{vmatrix} = (5-\lambda)(2-\lambda) - 4 = 0.$$

i.e. $\lambda^2 - 7\lambda + 6 = 0.$

$(\lambda - 1)(\lambda - 6) = 0 \quad \text{or} \quad \lambda = 1, 6.$

For $\lambda = 1$:
$$(A - \lambda I)e_1 = \begin{bmatrix} 4 & 4 \\ 1 & 1 \end{bmatrix} \begin{bmatrix} e_{11} \\ e_{12} \end{bmatrix} = 0.$$

i.e. $e_{11} + e_{12} = 0$ or $e_{11} = -e_{12}$.

Hence a possible characteristic vector is
$$\underset{\sim}{e_1} = \begin{bmatrix} 1 \\ -1 \end{bmatrix}.$$

For $\lambda = 6$: $\begin{bmatrix} -1 & 4 \\ 1 & -4 \end{bmatrix} \begin{bmatrix} e_{21} \\ e_{22} \end{bmatrix} = 0$; or $e_{21} = 4e_{22}$

Hence a possible vector is $\underset{\sim}{e_2} = \begin{bmatrix} 4 \\ 1 \end{bmatrix}$

The modal matrix is
$$P = \begin{bmatrix} e_1 & e_2 \end{bmatrix} = \begin{bmatrix} 1 & 4 \\ -1 & 1 \end{bmatrix}$$

and $P^{-1} = \dfrac{\text{adj } P}{|P|} = \dfrac{1}{5} \begin{bmatrix} 1 & -4 \\ 1 & 1 \end{bmatrix}$

Since $P^{-1} A P = \begin{bmatrix} 1 & 0 \\ 0 & 6 \end{bmatrix} = \Lambda$

$$e^{At} = Pe^{\Lambda t}P^{-1}$$

$$= \begin{bmatrix} 1 & 4 \\ -1 & 1 \end{bmatrix} \begin{bmatrix} e^t & 0 \\ 0 & e^{6t} \end{bmatrix} \begin{bmatrix} \tfrac{1}{5} & -\tfrac{4}{5} \\ \tfrac{1}{5} & \tfrac{1}{5} \end{bmatrix}$$

$$= \begin{bmatrix} 1 & 4 \\ -1 & 1 \end{bmatrix} \begin{bmatrix} \tfrac{1}{5}e^t & -\tfrac{4}{5}e^t \\ \tfrac{1}{5}e^{6t} & \tfrac{1}{5}e^{6t} \end{bmatrix}$$

$$= \begin{bmatrix} \left(\tfrac{1}{5}e^t + \tfrac{4}{5}e^{6t}\right) & \left(-\tfrac{4}{5}e^t + \tfrac{4}{5}e^{6t}\right) \\ \left(-\tfrac{1}{5}e^t + \tfrac{1}{5}e^{6t}\right) & \left(\tfrac{4}{5}e^t + \tfrac{1}{5}e^{6t}\right) \end{bmatrix}$$

9-2

Find a set of state equations and the state transition matrix for the circuit shown. Use i and v as shown for the states.

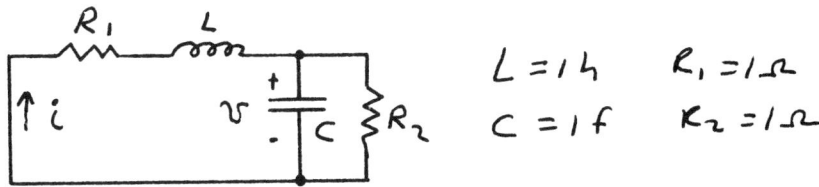

$L = 1 h \qquad R_1 = 1 \Omega$
$C = 1 f \qquad R_2 = 1 \Omega$

****

let $x_1 = i$, $x_2 = v$

$L\dot{x}_1 = -x_2 - R_1 x_1 \longrightarrow \dot{x}_1 = -x_2 - x_1$

$C\dot{x}_2 = -x_2/R_2 + x_1 \longrightarrow \dot{x}_2 = -x_2 + x_1$

$$\boxed{\underline{\dot{x}} = \begin{bmatrix} -1 & -1 \\ 1 & -1 \end{bmatrix} \underline{x} = A\underline{x}}$$ State equation

State transition matrix $= \phi(t) = e^{At}$

$$e^{At} = \mathcal{L}^{-1}[(sI-A)^{-1}] = \mathcal{L}^{-1}\left\{ \begin{bmatrix} s+1 & +1 \\ -1 & s+1 \end{bmatrix}^{-1} \right\}$$

$$= \mathcal{L}^{-1}\left\{ \frac{\begin{bmatrix} s+1 & -1 \\ +1 & s+1 \end{bmatrix}}{s^2 + 2s + 2} \right\} = \mathcal{L}^{-1}\begin{bmatrix} \frac{s+1}{(s+1)^2+1} & \frac{-1}{(s+1)^2+1} \\ \frac{+1}{(s+1)^2+1} & \frac{s+1}{(s+1)^2+1} \end{bmatrix}$$

$$\boxed{\phi(t) = e^{At} = \begin{bmatrix} e^{-t}\cos t & -e^{-t}\sin t \\ +e^{-t}\sin t & e^{-t}\cos t \end{bmatrix}}$$

10
FOURIER METHODS

THE TRIGONOMETRIC FOURIER SERIES

━━━ 10-1

Obtain the trigonometric Fourier series for v(t). Find the r.m.s. value of the first three terms of the series.

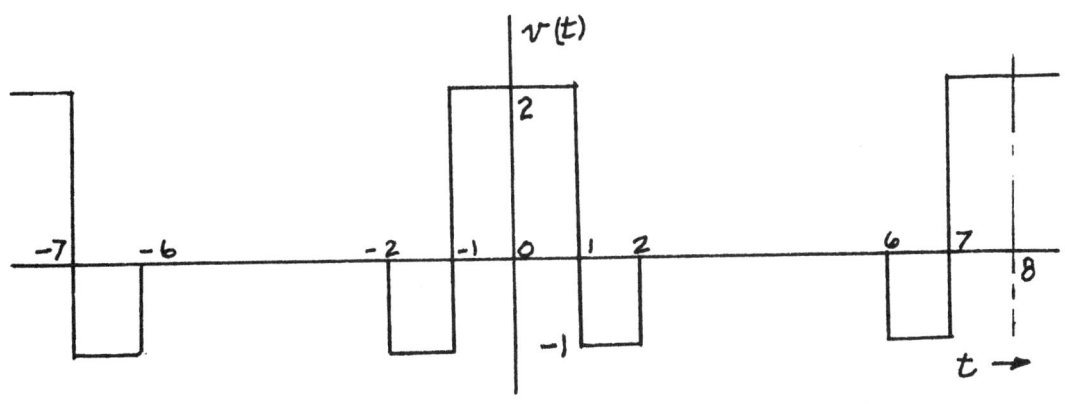

* * * * * * * * * * * * * * * * * * * *

Period $T = 8$ secs., $\omega_0 = \dfrac{2\pi}{T} = \dfrac{\pi}{4}$ rad/sec.

Even function, therefore $b_n = 0$

Form of series: $v(t) = \dfrac{a_0}{2} + \sum\limits_{n=1}^{\infty} a_n \cos n\omega_0 t$

$\dfrac{a_0}{2} = \dfrac{Area}{Period} = \dfrac{4-2}{8} = \dfrac{1}{4} \Rightarrow a_0 = \dfrac{1}{2}$

269

For $n \geq 1$,

$$a_n = \frac{4}{T} \int_0^{T/2} v(t) \cos n\omega_0 t \, dt = \frac{1}{2}\left[\int_0^1 2\cos\frac{n\pi}{4}t \, dt - \int_1^2 \cos\frac{n\pi}{4}t \, dt\right]$$

$$a_n = \frac{1}{2}\left[2\frac{\sin\frac{n\pi}{4}t}{\frac{\pi n}{4}}\Big|_0^1 - \frac{\sin\frac{n\pi}{4}t}{\frac{\pi n}{4}}\Big|_1^2\right]$$

$$a_n = \frac{2}{\pi n}\left[2\sin\frac{n\pi}{4} - \left(\sin\frac{n\pi}{2} - \sin\frac{n\pi}{4}\right)\right]$$

$$a_n = \frac{2}{\pi n}\left[3\sin\frac{n\pi}{4} - \sin\frac{n\pi}{2}\right]$$

Evaluating the first three non-dc coefficients,

$$a_1 = \frac{2}{\pi}\left[\frac{3}{\sqrt{2}} - 1\right] = 0.71385$$

$$a_2 = \frac{1}{\pi}\left[3\sin\frac{\pi}{2}\right] = 0.9549$$

$$a_3 = \frac{2}{3\pi}\left[\frac{3}{\sqrt{2}} + 1\right] = 0.6624$$

$$v(t) = \frac{1}{4} + a_1\cos\frac{\pi}{4}t + a_2\cos\frac{\pi}{2}t + a_3\cos\cos\frac{3\pi}{4}t + \ldots$$

The rms value of the first three terms is:

$$V_{rms}^2 = \left(\frac{1}{4}\right)^2 + \frac{1}{2}\left((0.7138)^2 + (0.9549)^2\right)$$

$$V_{rms}^2 = 0.0625 + 0.7107$$

$$V_{rms} = 0.8793 \text{ volts}$$

In the Fourier series for the periodic function below:
A. Are there any sine terms? Why?
B. Are there any cosine terms? Why?
C. Are there any even harmonics? Why?
D. What is the average D.C. level?

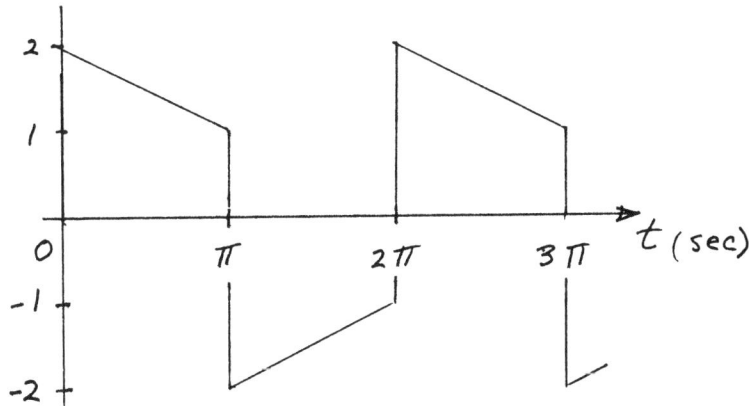

****

A. Yes, it is not an even function $f(t) \neq f(-t)$

B. Yes, it is not an odd function $f(t) \neq -f(-t)$

C. No, it is odd half wave $f(t) = -f\left(t + \frac{T}{2}\right)$

D. Zero, it is symmetrical about the t axis.

10-3

Determine the trigonometric Fourier series for the waveforms shown.

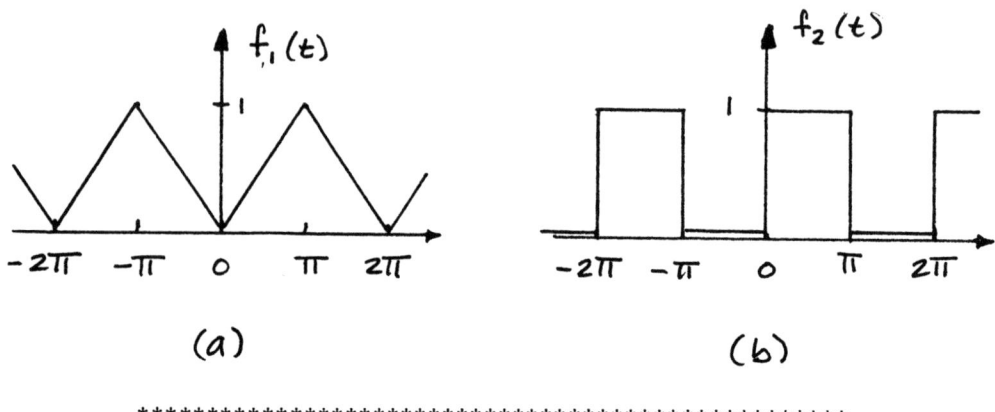

(a)   (b)

(a) $f_1(t)$ is even and symmetrical about $f_1(t)$ axis. By inspection, the average value is $1/2$ and the sine terms are all zero.

$$T = 2\pi, \quad \omega = \frac{2\pi}{T} = 1 \; \frac{rad}{sec}$$

$$a_N = \frac{2}{\pi} \int_0^\pi \frac{t}{\pi} \cos Nt \, dt$$

for N odd, $a_N = -\dfrac{4}{N^2 \pi^2}$

for N even, $a_N = 0$

$$\boxed{f_1(t) = \frac{1}{2} - \frac{4}{\pi^2}\left(\cos t + \frac{1}{9}\cos 3t + \frac{1}{25}\cos 5t + \cdots\right)}$$

10-4

(b) $f_2(t)$ is neither even nor odd. However, by removing the dc component (the average value), the function becomes an odd function.

$$b_0 = 1/2 \qquad f_2^*(t) = f_2(t) - 1/2$$

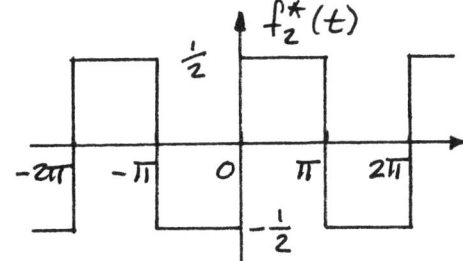

$$T = 2\pi$$
$$\omega = \frac{2\pi}{T} = 1 \; \frac{rad}{sec}$$

Since $f_2^*(t)$ is odd, the cosine terms are zero.

$$b_N = \frac{2}{\pi} \int_0^{\pi} \frac{1}{2} \sin Nt \, dt$$

for N odd, $b_N = \frac{2}{N\pi}$

for N even, $b_N = 0$

$$\boxed{f_2(t) = f_2^*(t) + \frac{1}{2} = \frac{1}{2} + \frac{2}{\pi}\left(\sin t + \frac{1}{3}\sin 3t + \frac{1}{5}\sin 5t + \cdots\right)}$$

10-5

The signal below is the output of a nonlinear device with a sinusoidal input. By first finding the Fourier series for this output determine the output of an ideal bandpass filter whose output only passes the 3rd harmonic. Find the optimum θ so that the filter output is a maximum and find this maximum.

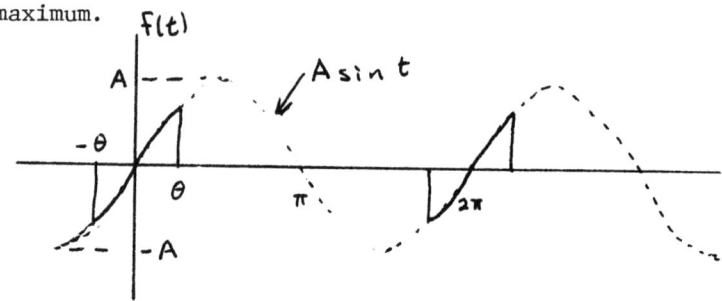

****

We have odd symmetry in this signal, so $a_n = 0$.

$$\therefore \quad f(t) = \sum_{n=1}^{\infty} b_n \sin nt$$

$$b_n = \frac{4}{T} \int_0^{T/2} f(t) \sin nt \, dt = \frac{2A}{\pi} \int_0^{\theta} \sin t \sin nt \, dt$$

For $n \neq 1$,

$$b_n = \frac{A}{\pi} \left[\frac{\sin(1-n)t}{1-n} - \frac{\sin(1+n)t}{1+n} \right]_0^{\theta}$$

$$= \frac{A}{\pi} \left[\frac{\sin(1-n)\theta}{1-n} - \frac{\sin(1+n)\theta}{1+n} \right]$$

$$b_1 = \frac{2A}{\pi} \int_0^{\theta} \sin^2 t \, dt = \frac{2A}{\pi} \left(\frac{1}{2}t - \frac{1}{4}\sin 2t \right)\bigg|_0^{\theta} = \frac{A}{\pi}(\theta - \frac{1}{2}\sin 2\theta)$$

$$f(t) = \frac{A}{\pi} \left\{ (\theta - \frac{1}{2}\sin 2\theta) + \sum_{n=2}^{\infty} \left[\frac{\sin(1-n)\theta}{1-n} - \frac{\sin(1+n)\theta}{1+n} \right] \sin nt \right\}$$

The filter output is the term for $n = 3$, whose magnitude is
$$b_3 = \frac{A}{\pi} \left[\frac{\sin(-2\theta)}{-2} - \frac{\sin 4\theta}{4} \right]$$

$$b_3 = \frac{A}{4\pi}(2\sin 2\theta - \sin 4\theta)$$

This becomes a maximum when $\frac{db_3}{d\theta} = 0$,

$$\frac{db_3}{d\theta} = \frac{A}{4\pi}(4\cos 2\theta - 4\cos 4\theta) = 0$$

$$\cos 2\theta = \cos 4\theta = 2\cos^2 2\theta - 1$$

$$2\cos^2 2\theta - \cos 2\theta - 1 = 0$$

$$\cos 2\theta = \frac{-1 \pm \sqrt{1+8}}{4} = \frac{-1 \pm 3}{4} = -1, \frac{1}{2}$$

$\cos 2\theta = -1$	$\cos 2\theta = \frac{1}{2}$
$2\theta = \pi$	$2\theta = \pi/3$
$\theta = \pi/2$	$\theta = \pi/6$

$$b_3\left(\frac{\pi}{2}\right) = \frac{A}{4\pi}(2\sin\pi - \sin 2\pi) = 0$$

$$b_3\left(\frac{\pi}{6}\right) = \frac{A}{4\pi}\left(2\sin\frac{\pi}{3} - \sin\frac{2\pi}{3}\right) =$$

$$\frac{A}{4\pi}\left(2\frac{\sqrt{3}}{2} - \frac{\sqrt{3}}{2}\right) = \frac{A\sqrt{3}}{8\pi}$$

$$b_{3\,max} = 0.0689\,A \quad \text{at} \quad \theta = \pi/6 = 30°$$

10-6

Find the value of the Fourier coefficient, a_2, for the periodic voltage waveform, $v(t)$.

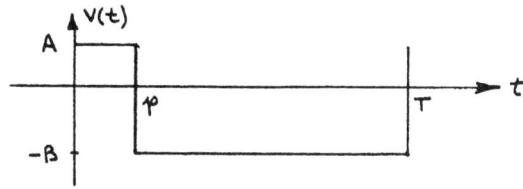

****

$$a_2 = \frac{2}{T}\int_0^T v(t)\cos 2\omega t\,dt = \frac{2}{T}\int_0^p A\cos 2\omega t\,dt + \frac{2}{T}\int_p^T (-B)\cos 2\omega t\,dt$$

$$= \frac{A+B}{2\pi}\sin\frac{4\pi}{T}p$$

SYMMETRY PROPERTIES

10-7

For the given periodic function, determine the trigonometric Fourier series coefficients, a_2 and b_2.

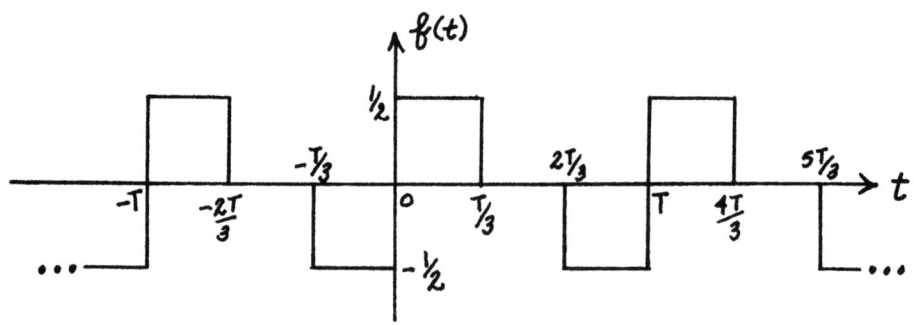

***

$f(t)$ is an odd function $\therefore a_n = 0 \; \forall n \Rightarrow \underline{a_2 = 0}$

$b_n = \dfrac{4}{T} \displaystyle\int_0^{T/2} f(t) \sin(n\omega_0 t)\, dt ; \quad \omega_0 = \dfrac{2\pi}{T}$

$b_2 = \dfrac{4}{T} \displaystyle\int_0^{T/3} (1/2) \sin(2\omega_0 t)\, dt = \left(\dfrac{2}{T}\right)\left(\dfrac{1}{2\omega_0}\right)\left[-\cos(2\omega_0 t)\right]_0^{T/3}$

$b_2 = \dfrac{1}{2\pi}\left(1 - \cos\dfrac{4\pi}{3}\right) = \underline{0.239}$

10-8

For the periodic waveform shown:

A. Use symmetry to state as much as possible about the trigonometric Fourier series for the waveform.
B. What is the fundamental frequency?
C. Find the rms amplitudes of the fundamental and third harmonic.

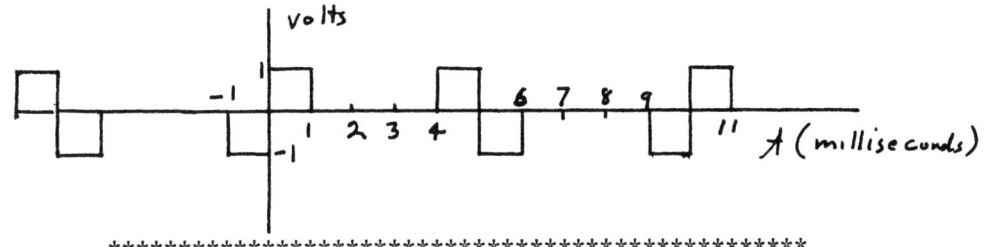

****

A. We see that the function repeats itself every 10 ms, and that there are equal areas above and below the axis. Hence, the dc term $\boxed{a_0 = 0}$

The waveform is an <u>odd</u> function of time ($f(t) = -f(-t)$) Hence, there are <u>no</u> cosine terms $\boxed{a_n = 0}$

The function has ½ wave symmetry (from 5 to 10 it is the negative of itself from 0 to 5. Hence, $b_n = 0$ for all <u>even</u> n.

B. $f = 1/T = 1/10 \times 10^{-3} \text{ sec} = 1000/10 = \boxed{100 \text{ Hz}}$

C. With half-wave symmetry, we need integrate over only one-half period. (Actually, the waveform has ¼ wave symmetry, meaning each half of the ½ waves chosen has an even symmetry about the midpoint of the half-wave.) We <u>could</u> integrate over ¼ period and multiply by 4, but we'll take ½ period.)

$$b_n = \frac{4}{10^{-2}} \left[\int_0^{10^{-3}} \sin 200\pi n t \, dt + \int_{4\times 10^{-3}}^{5\times 10^{-3}} \sin 200\pi n t \, dt \right]$$

$$= -\frac{400}{200\pi n} \left(\cos 200\pi n t \Big]_0^{10^{-3}} + \cos 200\pi n t \Big]_{4\times 10^{-3}}^{5\times 10^{-3}} \right)$$

$$= -\frac{2}{\pi n} \left(\cos(0.2\pi n) - \cos 0 + \cos \pi n - \cos(0.8\pi n) \right)$$

Fundamental: $b_1 = -\frac{2}{\pi} \left(\cos 0.2\pi - 1 + \cos \pi - \cos 0.8\pi \right)$

$= \frac{2}{\pi} (2 + \cos .8\pi - \cos .2\pi) = .243 \text{ v}$. However, this is the <u>peak</u> value of a sine wave. rms is .707 times as much. For fundamental, this is $\boxed{.172v}$

$b_3 = \frac{-2}{3\pi} (\cos 0.6\pi - 1 + \cos 3\pi - \cos 0.4\pi) = .555$

rms value of third harmonic is $.555/\sqrt{2} = \boxed{.393v}$

RESPONSE TO PERIODIC EXCITATIONS

━━━ 10-9

Given the following network and periodic input voltage $v_{in}(t)$, find the third harmonic of the output voltage $v_{out}(t)$.

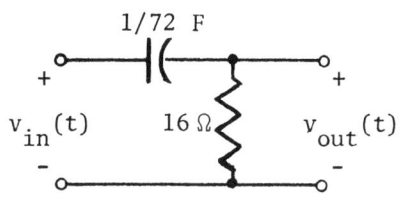
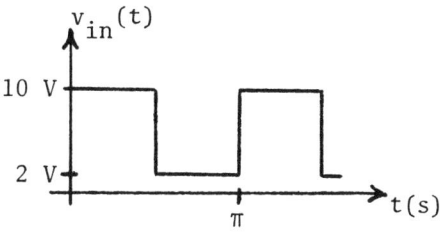

****

A downward shift of the input function to eliminate the dc component simplifies the work without affecting the input third harmonic. A downward shift by 6 V gives

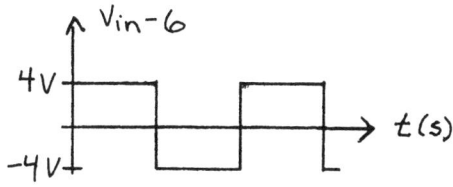

which is an odd function. It also has half-wave symmetry, but that property is not significant for determining the third harmonic of the output voltage. Since the shifted function is odd, its Fourier series has no cosine terms. So, we need only find b_3, the coefficient of the input third-harmonic sine term. In doing this we use the fact that $\omega_0 = 2\pi/T = 2\pi/\pi = 2$ rad/s, and that because of the symmetry, we need only integrate from 0 to T/2 if we multiply the integral by 2.

Proceeding,

$$b_3 = \frac{4}{T}\int_0^{T/2} f(t)\sin 3\omega_0 t\, dt = \frac{4}{\pi}\int_0^{\pi/2} 4\sin 6t\, dt$$

$$= -\frac{16}{6\pi}\cos 6t\Big|_0^{\pi/2} = -0.849\left(\cos\frac{6\pi}{2}-\cos 0\right)$$

Then, since $\cos 3\pi = -1$ and $\cos 0 = 1$

$$b_3 = -0.849(-2) = 1.698\ V$$

And the third harmonic term is $1.698\sin 6t$ V. The corresponding phasor is $1.698\ \underline{/-90°}$ V. At the third harmonic frequency of 6 rad/s, the network transfer function is

$$\frac{16}{16-j72/6} = \frac{16}{16-j12} = 0.8\ \underline{/36.9°}$$

So, the phasor for the third-harmonic of V_{out} is

$$(0.8\ \underline{/36.9°})(1.698\ \underline{/-90°}) = 1.358\ \underline{/-53.1°}\ V$$

The corresponding sinusoid is $1.358\cos(6t-53.1°)$ V, which is the output third-harmonic term.

10-10

Find the Fourier series of v(t). The input signal is periodic.

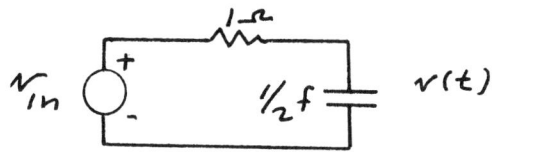
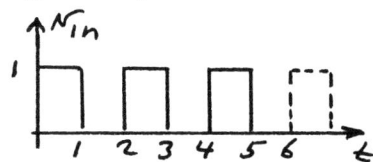

****

$$V_{in} = \frac{a_0}{2} + \sum_{n=1}^{\infty}(a_n \cos n\omega_0 t + b_n \sin n\omega_0 t) \qquad \omega_0 = \frac{2\pi}{T} = \pi$$

$$a_0 = 2/T \int_0^T V_{in} dt = \frac{2}{2}\left[\int_0^1 1 \, dt + \int_1^2 0 \, dt\right] = 1$$

$$a_n = 2/T \int_0^T V_{in} \cos n\omega_0 t \, dt = \frac{2}{T}\left[\int_0^1 \cos n\pi t \, dt + \int_1^2 0 \, dt\right]$$

$$= \left.\frac{\sin n\pi t}{n\pi}\right|_0^1 = 0$$

$$b_n = 2/T \int_0^T V_{in} \sin n\omega_0 t \, dt = \frac{2}{2}\left[\int_0^1 \sin n\pi t \, dt + \int_1^2 0 \, dt\right]$$

$$= \left.\frac{-\cos n\pi t}{n\pi}\right|_0^1 = \frac{2}{n\pi} \; n \text{ odd}, = 0 \; n \text{ even}$$

$$V_{in} = \frac{1}{2} + \sum_{\ell=1}^{\infty} \frac{2}{(2\ell-1)\pi} \sin\left[(2\ell-1)\pi t\right] \qquad \left\{\begin{array}{l}\text{let} \\ n = 2\ell-1\end{array}\right\}$$

By phasor method

$$\frac{V}{V_{in}}(j\omega) = \frac{1/j\omega c}{R + 1/j\omega c} = \frac{1}{j\omega RC+1} = \frac{1}{\sqrt{\omega^2 R^2 C^2 + 1}} \underline{/-\tan^{-1}\omega RC}$$

$$\frac{V}{V_{in}}(j(2\ell-1)\pi) = \frac{1}{\sqrt{(2\ell-1)^2\pi^2 \cdot \frac{1}{4} + 1}} \underline{/-\tan^{-1}\left[\frac{(2\ell-1)\pi}{2}\right]}$$

$$v(t) = \frac{1}{2} + \sum_{\ell=1}^{\infty} \frac{2}{(2\ell-1)\pi} \frac{\sin\left[(2\ell-1)\pi t - \tan^{-1}\left[\frac{(2\ell-1)\pi}{2}\right]\right]}{\sqrt{\frac{(2\ell-1)^2\pi^2}{4} + 1}}$$

COMPLEX FORM OF THE FOURIER SERIES

10-11

For the waveform shown, determine the complex form of the Fourier series.

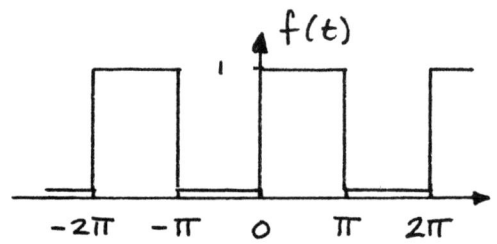

Redrawing the waveform so as to center it about its average value of $\frac{1}{2}$,

$$T = 2\pi$$
$$\omega = \frac{2\pi}{T} = 1 \text{ rad/sec}$$

$$C_N = \frac{1}{T}\int_{-T/2}^{T/2} f^*(t) e^{-jN\omega t} dt = \frac{1}{2\pi}\left[\int_{-\pi/2}^{\pi/2} f^*(t) e^{-jNt} dt\right]$$

$$C_N = \frac{1}{2\pi}\left[\int_{-\pi/2}^{0} -\tfrac{1}{2} e^{-jNt} dt + \int_{0}^{\pi/2} \tfrac{1}{2} e^{-jNt} dt\right]$$

$$C_N = -j\frac{1}{\pi}\int_{0}^{\pi}\tfrac{1}{2}\sin Nt\, dt$$

for N odd, $C_N = -j\frac{1}{N\pi}$

for N even, $C_N = 0$

So $f(t) = f^*(t) + \tfrac{1}{2} = \tfrac{1}{2} + \sum_{n=-\infty}^{\infty} \frac{-j e^{-j(2n+1)t}}{(2n+1)\pi}$

10-12

A periodic function f(t) consists of rising and decaying exponentials of time-constants of 1/5 s each and durations of 1 s each. Obtain the complex Fourier coefficients of f(t).
Assume $e^{-5} = 0$.

****

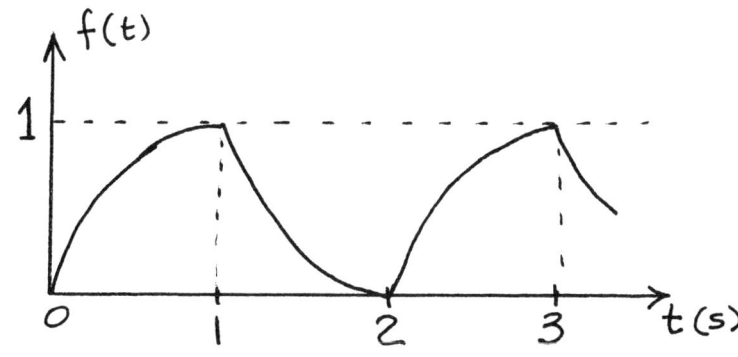

The period is $T = 2s$ and $\omega_o = \frac{2\pi}{2} = \pi$ rads/sec.

Within the first period, $f(t)$ can be expressed as

$$f(t) = \begin{cases} 1 - e^{-5t}, & 0 < t < 1 \\ e^{-5(t-1)}, & 1 < t < 2 \end{cases}$$

The complex form of the Fourier series expansion of $f(t)$ is

$$f(t) = \sum_{n=-\infty}^{\infty} c_n e^{jn\pi t}$$

where

$$c_n = \frac{1}{T} \int_0^T f(t) e^{-jn\pi t} dt$$

$$= \frac{1}{2} \left[\int_0^1 (1 - e^{-5t}) e^{-jn\pi t} dt + \int_1^2 e^{-5(t-1)} e^{-jn\pi t} dt \right]$$

$$= \frac{1}{2} \left[\int_0^1 e^{-jn\pi t} dt - \int_0^1 e^{-(5+jn\pi)t} dt + e^5 \int_1^2 e^{-(5+jn\pi)t} dt \right]$$

$$= \frac{1}{2} \left[\left. \frac{e^{-jn\pi t}}{-jn\pi} \right|_0^1 - \left. \frac{e^{-(5+jn\pi)t}}{-(5+jn\pi)} \right|_0^1 + \left. \frac{e^5 \, e^{-(5+jn\pi)t}}{-(5+jn\pi)} \right|_1^2 \right]$$

$$= \frac{1}{2} \left[\frac{e^{-jn\pi} - e^0}{-jn\pi} + \frac{e^{-(5+jn\pi)} - e^0}{5+jn\pi} - \frac{e^5 \left(e^{-2(5+jn\pi)} - e^{-(5+jn\pi)} \right)}{5+jn\pi} \right]$$

Using the relations $e^0 = 1$

$$e^{-jn\pi} = \cos n\pi = (-1)^n$$

we have   and $e^{-5} \approx 0$

$$c_n = \frac{1}{2} \left[\frac{1-(-1)^n}{jn\pi} - \frac{1}{5+jn\pi} + \frac{(-1)^n}{5+jn\pi} \right]$$

$$= \frac{1}{2} \left[1 - (-1)^n \right] \left[\frac{1}{jn\pi} - \frac{1}{5+jn\pi} \right]$$

$$= \frac{5}{(jn\pi)(5+jn\pi)}, \quad n = 1, 3, 5, \ldots$$

THE FOURIER TRANSFORM

━━ 10-13

Using the defining integral, determine the Fourier transform (in rectangular form) for the function given below.

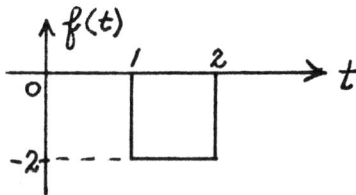

****

$$F(j\omega) = \int_{-\infty}^{\infty} f(t) e^{-j\omega t} dt = \int_{1}^{2} (-2) e^{-j\omega t} dt = \frac{2}{j\omega} \left[e^{-j\omega t} \right]_{1}^{2}$$

$$F(j\omega) = \frac{2}{\omega} (\sin\omega - \sin 2\omega) + j\frac{2}{\omega}(\cos\omega - \cos 2\omega)$$

10-14

Use Parseval's theorem to find the cutoff frequency of the ideal low pass filter so that the energy in x(t) is the same as the energy in y(t).

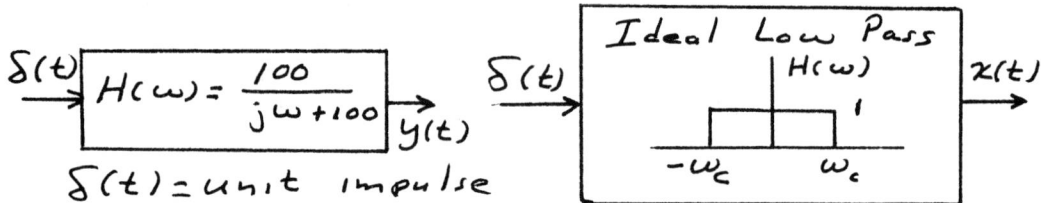

$\delta(t)$ = unit impulse

***

$\mathcal{F}(\delta(t)) = 1 \rightarrow Y(\omega) = \dfrac{100}{j\omega + 100}$ and $X(\omega) = 1 \ |\omega| < \omega_c$
$= 0 \ |\omega| > \omega_c$

$$E_y = \frac{1}{2\pi} \int_{-\infty}^{\infty} Y(\omega) Y(-\omega) \, d\omega = \text{Energy}$$

$$= \frac{1}{2\pi} \int_{-\infty}^{\infty} \left(\frac{100}{j\omega + 100}\right)\left(\frac{100}{-j\omega + 100}\right) d\omega$$

$$= \frac{1}{2\pi} \int_{-\infty}^{\infty} \frac{10^4}{\omega^2 + 100^2} d\omega = \frac{1}{2\pi} \frac{10^4}{100} \tan^{-1}\frac{\omega}{100} \Big|_{-\infty}^{\infty}$$

$$= \frac{1}{2\pi} 10^2 [\pi/2 - (-\pi/2)] = \boxed{50}$$

$$E_x = \frac{1}{2\pi} \int_{-\infty}^{\infty} X(\omega) X(-\omega) \, d\omega$$

$$= \frac{1}{2\pi} \int_{-\omega_c}^{\omega_c} 1 \, d\omega = \frac{\omega_c}{\pi} = E_y = 50$$

$$\boxed{\omega_c = 50\pi \text{ rad/sec}}$$

11
LAPLACE TRANSFORMS

IMPULSE RESPONSE AND CONVOLUTION

--11-1

For the circuit shown:
A. Find the transfer function relating v_2 to v_1
B. The unit impulse response of the circuit
C. From B, sketch v_2 for 2 msec after a pulse of 4 volts amplitude and 0.2 msec duration is applied as v_1

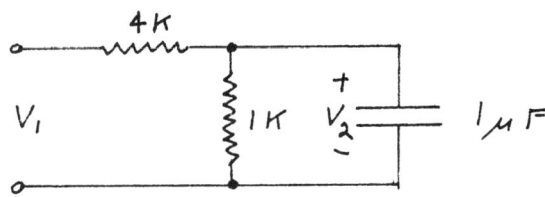

****

A.
Transfer impedance of the parallel R and C is

$$\frac{R\left(\frac{1}{sc}\right)}{R + \frac{1}{sc}} = \frac{R}{1 + sCR} = \frac{10^3}{1 + s\,10^{-6} \times 10^3} = \frac{10^3}{1 + 10^{-3}s}$$

Now we apply the voltage-division theorem to get $H(s)$

$$H(s) = \frac{\frac{1000}{1+10^{-3}s}}{4000 + \frac{1000}{1+10^{-3}s}} = \frac{1000}{4000+4s+1000} = \frac{1}{4}\cdot\frac{1000}{s+1250} = \frac{250}{s+1250}$$

B. The inverse transform of $H(s)$ is $h(t) = 250 e^{-1250 t}$

C. $h(t)$ gives the response to a pulse for which amplitude × duration is 1 volt-second. For the pulse stated, we have $4v \times 2\times10^{-4}\text{sec} = 8\times10^{-4}$ v-sec, and after the pulse ends, the response will simply be $8\times10^{-4} h(t) = 0.2 e^{-1250 t}$.

Because pulse duration is small compared to time constant (We are about at the limit where that assumption can be made with reasonable accuracy) during the pulse, voltage rises almost linearly in time to 0.2v. If we are required to prove this, we may write from experience that during the 4v pulse, voltage will rise exponentially toward the value determined by the voltage divider $\frac{1K}{4K+1K}\times 4v = 0.8v$

The Thevenin resistance of the circuit charging C is $4K \parallel 1K = \frac{4K \times 1K}{5K} = 800\Omega$. Hence the time constant $RC = 800 \times 10^{-6} = 0.8$ msec, so $\frac{1}{\tau} = 1250$ sec^{-1}

$v = 0.8(1 - e^{-1250 t}) \simeq 0.8[1-(1-1250 t)] = 0.8 \times 1250 \times 2\times10^{-4}$
$= 0.2 v$ (Solved exactly, we get .181v, which shows us the pulse was a <u>bit</u> long to be considered impulsive)

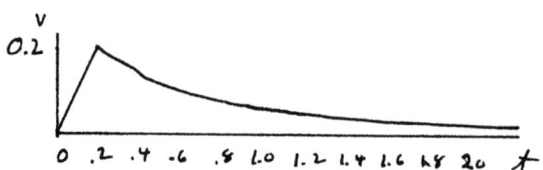

11-2

Find the impulse response of the circuit shown.

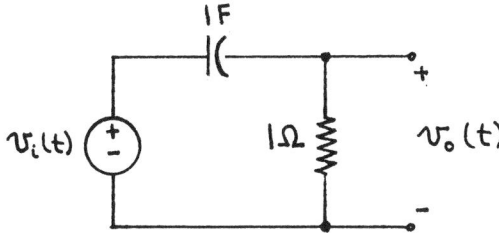

***

Divide the problem into two parts: one for a small time interval about $t=0$, which is much smaller than the circuit time constant ($\tau = RC = 1$ second), and the other for $(0, \infty)$. The equivalent circuit for the interval about $t=0$ is

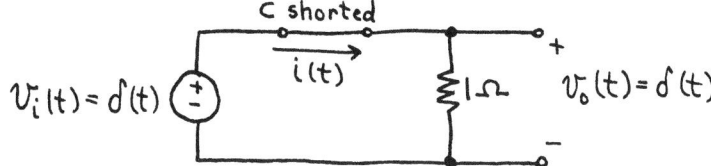

Now $i(t) = \dfrac{v_i(t)}{1\Omega} = \delta(t)$. The capacitor voltage jumps to

$$v(0^+) = \frac{1}{C}\int_{-\infty}^{0^+} \delta(t)\,dt = \frac{1}{C} = 1 \text{ volt},$$ with the positive sign to the left.

The circuit for $t > 0$ is

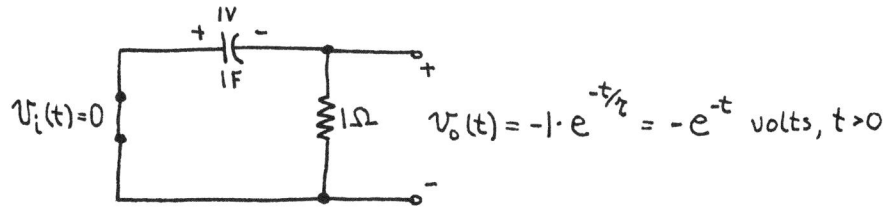

$v_o(t) = -1 \cdot e^{-t/\tau} = -e^{-t}$ volts, $t > 0$

Putting both pieces together, we have

$$h(t) = v_o(t) = \delta(t) - e^{-t}u(t)$$

11-3

Compute and sketch the convolution x(t) * y(t) of the two time functions shown.

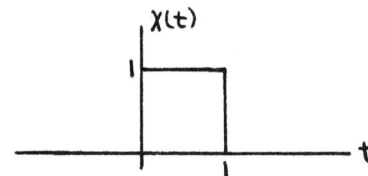 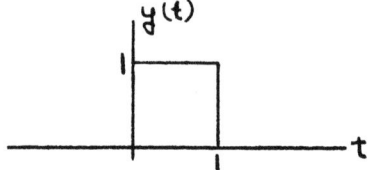

****

① $t < 0$: 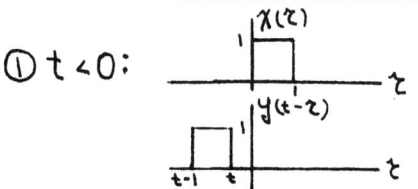 $\quad x(\tau)y(t-\tau)=0$ so $x(t)*y(t)=0$ for $t<0$

② $0 \leq t < 1$: $\quad x(t)*y(t) = \int_0^t 1\,d\tau = t$

③ $1 \leq t < 2$: $\quad x(t)*y(t) = \int_{t-1}^1 1\,d\tau$

$\qquad = 1-(t-1) = 2-t$

④ $2 \leq t$ 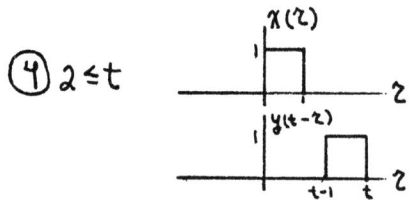 $\quad x(\tau)y(t-\tau)=0$ so $x(t)*y(t)=0$

Therefore:

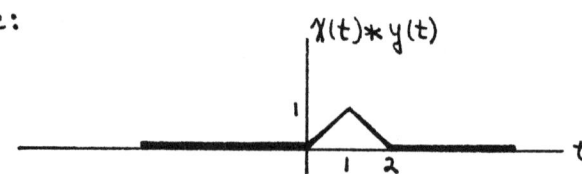

11-4

Find the impulse response of the circuit shown.

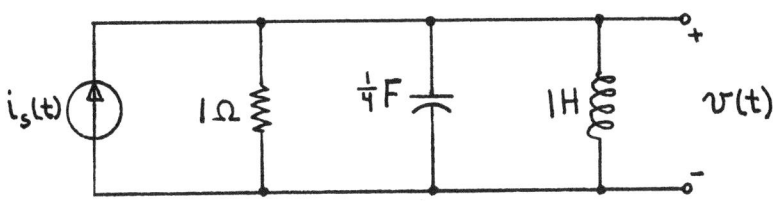

****
We break the problem into two parts — the response in an infinitesimal interval about $t=0$, the other for $t>0$. For the first, the circuit becomes (letting $i_s(t)=\delta(t)$):

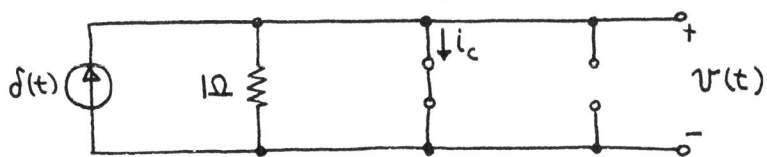

Note that the capacitor has no charge and the inductor no current at $t=0^-$; thus, they are a short circuit and an open circuit, respectively. Hence, $v(t)=0$; however, $i_c(t)=\delta(t)$. Therefore, $v(0^+)=\frac{1}{C}\int_{-\infty}^{0^+} i_c(t)dt = 4$ volts, with the + sign on top. Since $v(t)$ remains finite, the inductor current does not jump. The circuit for $t>0$ is

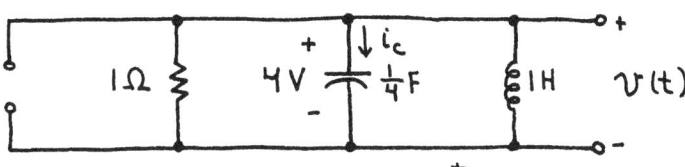

Nodal analysis gives $\frac{v}{1}+\frac{1}{4}\frac{dv}{dt}+\frac{1}{1}\int_{-\infty}^{t} v(x)dx = 0$.

Differentiating, $\frac{dv}{dt}+\frac{1}{4}\frac{d^2v}{dt^2}+v=0$, or $\frac{d^2v}{dt^2}+4\frac{dv}{dt}+4v=0$.

The characteristic equation is $s^2+4s+4=0$, or $(s+2)^2=0$.

Hence, the response is $v(t) = K_1 e^{-2t} + K_2 t e^{-2t}$. $v(0^+)=4=K_1$.

To get K_2, compute $\frac{dv}{dt}\big|_{0^+} = -2K_1 + K_2$. Now $i_c(0^+) = \frac{1}{4}\frac{dv}{dt}\big|_{0^+}$

$= \frac{1}{4}(-2 \times 4 + K_2) = -2 + \frac{1}{4}K_2$. What is $i_c(0^+)$? Drawing the circuit at $t = 0^+$:

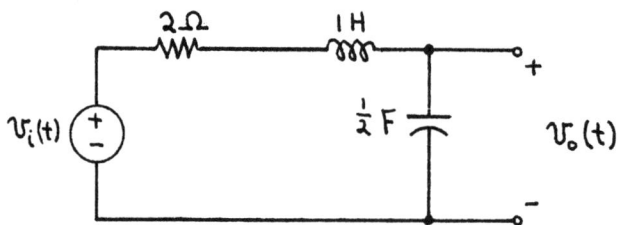

$i_c(0^+) = -4A$. Thus
$-4A = -2 + \frac{1}{4}K_2 \Rightarrow \frac{1}{4}K_2 = -2 \Rightarrow K_2 = -8$.

Thus, $h(t) = v(t) = (4e^{-2t} - 8te^{-2t})U(t)$.

DEFINITION OF THE LAPLACE TRANSFORM AND THE SYSTEM FUNCTION

11-5

Use LaPlace Transform techniques to find the impulse response of the circuit shown.

****

With all initial conditions zero, the S domain equivalent circuit becomes (with $v_i(t) = \delta(t)$):

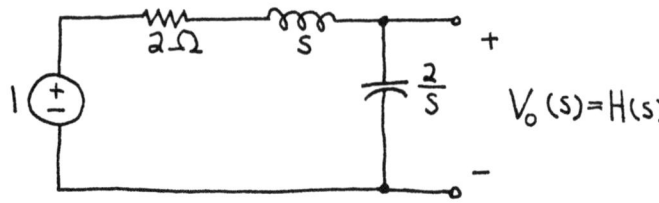

$$H(s) = \frac{\frac{2}{5}}{2+s+\frac{2}{5}} = \frac{2}{s^2+2s+2} = \frac{2}{(s+1)^2+1}$$

Thus, $h(t) = \mathcal{L}^{-1}\{H(s)\} = 2\mathcal{L}^{-1}\left\{\frac{1}{(s+1)^2+1}\right\}$

$$= 2e^{-t}\sin(t)U(t)$$

11-6

$2\delta(t) \longrightarrow$ [Linear System] $\longrightarrow 4e^{-2t}\cos(4t)U(t)$

Compute the system function $H(s) = \mathcal{L}\{h(t)\}$, and determine its region of convergence.

****

$h(t) = 2e^{-2t}\cos(4t)U(t)$ (must divide by 2)

$$H(s) = \int_{-\infty}^{\infty} h(t)e^{-st}dt = 2\int_{0}^{\infty} e^{-2t}\cos(4t)e^{-st}dt$$

$$= 2\int_{0}^{\infty} e^{-2t}\left[\frac{e^{j4t}+e^{-j4t}}{2}\right]e^{-st}dt = \int_{0}^{\infty} e^{-(s+2-j4)t}dt + \int_{0}^{\infty} e^{-(s+2+j4)t}dt$$

$$= \left.\frac{e^{-(s+2-j4)t}}{-(s+2-j4)}\right|_{0}^{\infty} + \left.\frac{e^{-(s+2+j4)t}}{-(s+2+j4)}\right|_{0}^{\infty}$$

Now if $\text{Re}(s) = \text{Re}(\sigma+j\omega) = \sigma > -2$, both upper limits are zero, and

$$H(s) = \frac{1}{s+2-j4} + \frac{1}{s+2+j4} = \frac{(s+2+j4)+(s+2-j4)}{(s+2)^2+16},$$

or $H(s) = 2\dfrac{s+2}{s^2+4s+20}$; $\text{Re}(s) = \sigma > -2$.

11-7

Find the impulse response of the circuit shown by using the LaPlace transform.

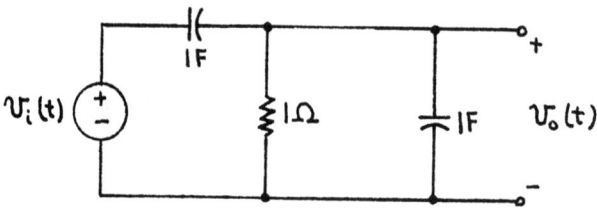

****

Find $H(s)$:

$$H(s) = \frac{1 \times \frac{1}{s}}{1 + \frac{1}{s}} \bigg/ \left(\frac{1}{s} + \frac{1 \times \frac{1}{s}}{1+\frac{1}{s}}\right) = \frac{\frac{1}{s+1}}{\frac{1}{s} + \frac{1}{s+1}} = \frac{s}{s+1+s} = \frac{s}{2s+1} = \frac{\frac{1}{2}s}{s+\frac{1}{2}}$$

Now $\mathcal{L}^{-1}\left\{\frac{1}{s+\frac{1}{2}}\right\} = e^{-\frac{1}{2}t}\, u(t)$

$\mathcal{L}^{-1}\left\{s \cdot \frac{1}{s+\frac{1}{2}}\right\} = \frac{d}{dt}\left[e^{-\frac{1}{2}t} u(t)\right] = -\frac{1}{2} e^{-\frac{1}{2}t} u(t) + (e^{-\frac{1}{2}t})\frac{d}{dt} u(t)$

$= -\frac{1}{2} e^{-\frac{1}{2}t} u(t) + e^{-\frac{1}{2}t}\delta(t) = -\frac{1}{2}e^{-\frac{1}{2}t} u(t) + \delta(t)$

(Note that $e^{-\frac{1}{2}t}\delta(t) = e^{-\frac{1}{2}\cdot 0}\cdot \delta(t) = \delta(t)$.)

Thus, $h(t) = \mathcal{L}^{-1}\left\{\frac{1}{2}\frac{s}{s+\frac{1}{2}}\right\} = \frac{1}{2}\left[-\frac{1}{2}e^{-\frac{1}{2}t} u(t) + \delta(t)\right]$

or $h(t) = -\frac{1}{4} e^{-\frac{1}{2}t} u(t) + \frac{1}{2}\delta(t)$

BASIC TRANSFORMS AND PROPERTIES

━━━━━━━━━━━━━━━━━━━━━━━━━━━━━━━ 11-8

Compute the LaPlace Transform for the waveform shown.

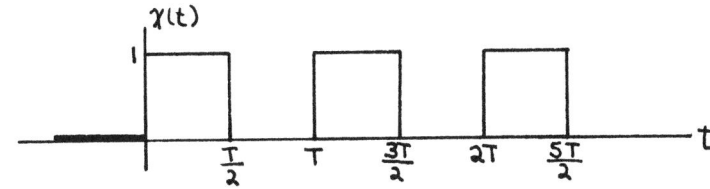

$$***$$

$$X(s) = \int_0^\infty x(t)e^{-st}dt = \sum_{k=0}^{\infty} \int_{kT}^{(k+1)T} x(t)e^{-st}dt$$

Change variable of integration

let $u = t - kT$, $du = dt$. Then $X(s) = \sum_{k=0}^{\infty} \int_0^T x(u+kT)e^{-s(u+kT)}du$

Now $x(t)$ is periodic with period T, so $x(u+kT) = x(u)$. Therefore,

$$X(s) = \sum_{k=0}^{\infty} \left[\int_0^T x(u)e^{-su}du\right] e^{-kTs}$$

$\int_0^T x(u)e^{-su}du = \int_0^{T/2} 1 \cdot e^{-su}du = \left.\frac{e^{-su}}{-s}\right|_0^{T/2} = \frac{1-e^{-\frac{sT}{2}}}{s}$, so

$$X(s) = \frac{1-e^{-\frac{sT}{2}}}{s} \cdot \sum_{k=0}^{\infty} (e^{-sT})^k.$$ If $\text{Re}(s) = \sigma > 0$,

then $|e^{-sT}| = |e^{-(\sigma+j\omega)T}| = e^{-\sigma T} < 1$, so $\sum_{k=0}^{\infty} (e^{-sT})^k = \frac{1}{1-e^{-sT}}$

(geometric series)

Finally, $$X(s) = \frac{1-e^{-\frac{sT}{2}}}{s(1-e^{-sT})}$$

11-9

Solve the differential equation

$$\frac{d^2y}{dt^2} + 2\frac{dy}{dt} + 2y = u(t)$$

using the LaPlace Transform.

****

We will use the LaPlace Transform $\mathcal{L}\{y(t)\} = \int_{0^-}^{\infty} y(t)e^{-st}\,dt$, since initial conditions are easier to handle with it. Using the differentiation property of the LaPlace Transform, $\mathcal{L}\{\frac{dy}{dt}\} = sY(s) - y(0^-)$, and $\mathcal{L}\{\frac{d^2y}{dt^2}\} = s\mathcal{L}\{\frac{dy}{dt}\} - \frac{dy}{dt}\big|_{0^-} = s^2Y(s) - sy(0^-) - \frac{dy}{dt}(0^-)$, we get

$$\left[s^2Y(s) - sy(0^-) - \frac{dy}{dt}(0^-)\right] + 2\left[sY(s) - y(0^-)\right] + Y(s) = \frac{1}{s}$$

Now --- if we assume that the solution is stable (goes to zero for large t) --- we can reason that $y(0^-) = \frac{dy}{dt}(0^-) = 0$, since $u(t) = 0$ for $t < 0$.

Thus,
$$(s^2 + 2s + 2)Y(s) = \frac{1}{s}, \text{ or } Y(s) = \frac{1}{s(s^2+2s+2)} = \frac{1}{s[(s+1)^2 + 1]}$$

By partial fractions, $Y(s) = \frac{K_0}{s} + \frac{K_1 s + K_2}{(s+1)^2 + 1} \cdot K_0 = [sY(s)]_{s \to 0} = \frac{1}{2} = K_0$

$$\therefore \frac{1}{s(s^2+2s+2)} = \frac{\frac{1}{2}}{s} + \frac{K_1 s + K_2}{s^2 + 2s + 2}, \text{ or } 2 = (s^2 + 2s + 2) + 2s(K_1 s + K_2).$$

Thus, $2 = (s^2 + 2s + 2) + (2K_1 s^2 + 2K_2 s) = (2K_1 + 1)s^2 + 2(K_2 + 1)s + 2$, or
$2K_1 + 1 = 0 \Rightarrow \underline{\underline{K_1 = -\frac{1}{2}}}$, $K_2 + 1 = 0 \Rightarrow \underline{\underline{K_2 = -1}}$

$$\therefore Y(s) = \frac{\frac{1}{2}}{s} + \frac{-\frac{1}{2}s - 1}{(s+1)^2 + 1} = \frac{\frac{1}{2}}{s} - \frac{1}{2}\frac{s}{(s+1)^2 + 1} - \frac{1}{(s+1)^2 + 1}$$

Inverting, we have $y(t) = \frac{1}{2}U(t) - \frac{1}{2}e^{-t}\cos(t)U(t) - e^{-t}\sin(t)U(t)$,

or $y(t) = \frac{1}{2}\left[1 - e^{-t}(\cos t + \sin t)\right]U(t)$

11-10

Prove that $\mathcal{L}\{f(t)e^{-at}\} = F(s+a)$ and use it to compute:
(a) $\mathcal{L}\{\cos(\omega_0 t)u(t)\}$
(b) $\mathcal{L}\{\sin(\omega_0 t)u(t)\}$
(c) $\mathcal{L}\{f(t)\cos(\omega_0 t)u(t)\}$

Note that $\mathcal{L}\{f(t)\} = F(s)$.

****

$$\mathcal{L}\{f(t)e^{-at}\} = \int_0^\infty f(t)e^{-at}\cdot e^{-st}\,dt = \int_0^\infty f(t)e^{-(s+a)t}\,dt = F(s+a)$$

(a) $\mathcal{L}\{u(t)\} = \frac{1}{s}$, so $\mathcal{L}\{\cos(\omega_0 t)u(t)\} = \mathcal{L}\left\{\frac{e^{j\omega_0 t} + e^{-j\omega_0 t}}{2}u(t)\right\}$

$$= \frac{1}{2}\mathcal{L}\{e^{j\omega_0 t}u(t)\} + \frac{1}{2}\mathcal{L}\{e^{-j\omega_0 t}u(t)\} = \frac{1}{2}\frac{1}{s-j\omega_0} + \frac{1}{2}\frac{1}{s+j\omega_0}$$

$$= \frac{1}{2}\frac{(s+j\omega_0)+(s-j\omega_0)}{s^2+\omega_0^2} = \frac{s}{s^2+\omega_0^2}.$$

(b) $\mathcal{L}\{\sin(\omega_0 t)u(t)\} = \mathcal{L}\left\{\frac{e^{j\omega_0 t} - e^{-j\omega_0 t}}{2j}u(t)\right\} = \frac{\mathcal{L}\{e^{j\omega_0 t}u(t)\} - \mathcal{L}\{e^{-j\omega_0 t}u(t)\}}{2j}$

$$= \frac{1}{2j}\left[\frac{1}{s-j\omega_0} - \frac{1}{s+j\omega_0}\right] = \frac{(s+j\omega_0)-(s-j\omega_0)}{2j(s^2+\omega_0^2)} = \frac{\omega_0}{s^2+\omega_0^2}.$$

(c) $\mathcal{L}\{f(t)\cos(\omega_0 t)u(t)\} = \mathcal{L}\left\{f(t)\frac{e^{j\omega_0 t}+e^{-j\omega_0 t}}{2}u(t)\right\}$

$$= \frac{1}{2}\mathcal{L}\{f(t)u(t)e^{j\omega_0 t}\} + \frac{1}{2}\mathcal{L}\{f(t)e^{-j\omega_0 t}u(t)\}$$

$$= \frac{1}{2}\mathcal{L}\{f(t)e^{j\omega_0 t}\} + \frac{1}{2}\mathcal{L}\{f(t)e^{-j\omega_0 t}\} = \frac{1}{2}F(s-j\omega_0) + \frac{1}{2}F(s+j\omega_0)$$

$$= \frac{F(s-j\omega_0) + F(s+j\omega_0)}{2}.$$

<u>Notes:</u> (1) In (a) & (b), the region of convergence of $\mathcal{L}\{u(t)\} = \frac{1}{s}$ is $\text{Re}(s) = \sigma > 0$, so this is the R.O.C. of $\mathcal{L}\{\cos(\omega_0 t)u(t)\}$ and $\mathcal{L}\{\sin(\omega_0 t)u(t)\}$.

(2) In (c), $f(t)u(t) = f(t)$, $t > 0$. The R.O.C. is the same as that for $F(s)$.

INVERSE TRANSFORMS
BY PARTIAL FRACTION EXPANSION

11-11

The current response of a network to a unit step of voltage is $i(t) = e^{-t} + 2e^{-2t}$ amps. What voltage, $v(t)$, must be applied to the network to obtain a response $i(t) = 2e^{-2t}$?

* *

For $V(t) = $ UNIT STEP THEN $V(s) = \dfrac{1}{s}$

For $i(t) = \varepsilon^{-t} + 2\varepsilon^{-2t}$ THEN $I(s) = \dfrac{1}{s+1} + \dfrac{2}{s+2}$

$$Y(s) = \dfrac{I(s)}{V(s)} = \dfrac{s(3s+4)}{(s+1)(s+2)}$$

For $i(t) = 2\varepsilon^{-2t}$, $I(s) = \dfrac{2}{s+2}$

$$V(s) = \dfrac{I(s)}{Y(s)} = \dfrac{2}{s+2} \cdot \dfrac{(s+1)(s+2)}{s(3s+4)}$$

$$V(s) = \dfrac{2(s+1)}{s(3s+4)} = \dfrac{K_1}{s} + \dfrac{K_2}{3s+4}$$

$K_1 = K_2 = \dfrac{1}{2}$ THUS $V(s) = \dfrac{1}{2} \cdot \left[\dfrac{1}{s} + \dfrac{1}{3s+4}\right] = \dfrac{1/2}{s} + \dfrac{1/6}{s+\frac{4}{3}}$

$$v(t) = \dfrac{1}{2}\left[1 + \dfrac{1}{3}\varepsilon^{-\frac{4}{3}t}\right]$$

11-12

a) Find h(t) for the system function below by partial fraction expansions and inverse Laplace transforms.

b) Verify that the initial value and final value theorems work for this function.

$$H(s) = \frac{2}{s(s+1)^2(s+2)}$$

****

a)
$$\frac{2}{s(s+1)^2(s+2)} = \frac{A}{s} + \frac{B}{(s+1)^2} + \frac{C}{s+1} + \frac{D}{s+2}$$

$$A = \frac{2}{(s+1)^2(s+2)}\bigg|_{s=0} = 1 \qquad B = \frac{2}{s(s+2)}\bigg|_{s=-1} = -2$$

$$D = \frac{2}{s(s+1)^2}\bigg|_{s=-2} = -1$$

C may be found by any of several methods. One method is to remove the 2nd order pole

$$\frac{A}{s} + \frac{C}{s+1} + \frac{D}{s+2} = \frac{2}{s(s+1)^2(s+2)} - \frac{B}{(s+1)^2} = \frac{2}{s(s+2)}$$

$$C = \frac{2(s+1)}{s(s+2)}\bigg|_{s=-1} = 0$$

$$\boxed{h(t) = (1 - 2te^{-t} - e^{-2t})u(t)}$$

b) $h(0+) = 1 - 0 - 1 = 0 = \lim_{s \to \infty} \frac{2s}{s(s+1)^2(s+2)} = 0$ ✓

$h(\infty) = 1 = \lim_{s \to 0} \frac{2s}{s(s+1)^2(s+2)} = 1$ ✓

11-13

Determine, by use of partial fractions, the inverse Laplace transform of the function shown below.

$$H(s) = \frac{s+3}{(s+1)(s^2+2s+5)}$$

$$H(s) = \frac{s+3}{(s+1)(s^2+2s+5)} = \frac{A}{s+1} + \frac{Bs+C}{s^2+2s+5}$$

$$A = H(s) \cdot (s+1) \Big|_{s=-1} = \frac{s+3}{s^2+2s+5}\Big|_{s=-1} = \frac{1}{2}$$

To find B and C, multiply both sides by $(s+1)(s^2+2s+5)$, collect terms, and equate like coefficients from both sides.

$$s+3 = A(s^2+2s+5) + (Bs+C)(s+1)$$

$$s+3 = (A+B)s^2 + (2A+B+C)s + (5A+C)$$

with $A = \frac{1}{2}$,

$$(s+3) = (B+\tfrac{1}{2})s^2 + (B+C+1)s + (C+\tfrac{5}{2})$$

$$\left.\begin{array}{l} B+\tfrac{1}{2} = 0 \\ B+C+1 = 1 \\ C+\tfrac{5}{2} = 3 \end{array}\right\} \Rightarrow \begin{array}{l} B = -\tfrac{1}{2} \\ C = \tfrac{1}{2} \end{array}$$

Therefore,

$$H(s) = \frac{1/2}{s} + \frac{-\frac{1}{2}s + \frac{1}{2}}{s^2 + 2s + 5}.$$ Completing the square on the second term puts it into a trigonometric form.

$$H(s) = \frac{1/2}{s} + \frac{-\frac{1}{2}s}{(s+1)^2 + 2^2} + \frac{\frac{1}{2}}{(s+1)^2 + 2^2}$$

$$= \frac{1/2}{s} - \frac{1}{2}\left[\frac{s-1}{(s+1)^2 + 2^2}\right]$$

$$H(s) = \frac{1/2}{s} - \frac{1}{2}\left[\frac{s+1}{(s+1)^2 + 2^2} - \frac{2}{(s+1)^2 + 2^2}\right]$$

$$h(t) = \left\{\frac{1}{2} - \frac{1}{2}\left[e^{-t}\cos 2t - e^{-t}\sin 2t\right]\right\}u(t)$$

$$\boxed{h(t) = \frac{1}{2} - \frac{1}{2}e^{-t}\left(\cos 2t - \sin 2t\right)}$$

11-14

Make a partial fraction expansion and take the inverse Laplace transform to find the response $v_o(t)$ to a unit impulse input $v_i(t)$ for this transfer function, $T(s)$.

$$\frac{V_o}{V_i} = T(s) = \frac{s^2 - 2s + 5}{s^2 + 2s + 5}$$

****

$$v_o(t) = \mathcal{L}^{-1}\left(\frac{s^2-2s+5}{s^2+2s+5}\right)\left(\mathcal{L}\,\delta(t)\right)$$

$$\mathcal{L}\,\delta(t) = 1$$

$$\frac{s^2-2s+5}{s^2+2s+5} = 1 + \frac{k_1 s + k_2}{(s+1)^2 + 2^2}$$

Since the function is 1 as $s \to \infty$ the first term in the partial fraction expansion is 1.

$$s^2 - 2s + 5 = s^2 + 2s + 5 + k_1 s + k_2$$

therefore $k_1 = -4$, $k_2 = 0$

$$v_o(t) = \mathcal{L}^{-1}\left\{1 - 4\left[\frac{s+1}{(s+1)^2 + 2^2}\right] + \frac{4}{2}\left(\frac{2}{(s+1)^2 + 2^2}\right)\right\}$$

$$\underline{v_o(t) = \delta(t) - 4e^{-t}\cos 2t + 2e^{-t}\sin 2t}$$

11-15

Determine the inverse Laplace transform of the following function.

$$I(s) = \frac{9s + 18}{s(3s^2 + 18s + 39)}$$

***

$$I(s) = \frac{3(s+2)}{s(s^2+6s+13)} = \frac{3(s+2)}{s(s+3-j2)(s+3+j2)}$$

$$I(s) = \frac{K_1}{s} + \frac{K_2}{s+3-j2} + \frac{K_2^*}{s+3+j2}$$

$$K_1 = \left. \frac{3(s+2)}{s^2+6s+13} \right|_{s=0} = 0.462$$

$$K_2 = \left. \frac{3(s+2)}{s(s+3+j2)} \right|_{s=-3+j2} = \frac{3(-3+j2+2)}{(-3+j2)(-3+j2+3+j2)} = 0.47 \angle{-119.7°}$$

$$I(s) = \frac{0.462}{s} + \frac{0.47\angle{-119.7°}}{s+3-j2} + \frac{0.47\angle{119.7°}}{s+3+j2}$$

$$\Rightarrow i(t) = 0.462 + (2)(0.47)e^{-3t}\cos(2t - 119.7°) \,;\, t \geq 0$$

11-16

Find the inverse LaPlace transform of $\dfrac{s + 20}{s^2 + 16s + 100}$

****

It is first necessary to find the poles of the function, i.e., the roots of the polynomial in the denominator. We could do this by applying the quadratic formula, but let's do it by completing the square. We use up $s^2 + 16s$ in a quantity of form:

$(s + \alpha)^2 = s^2 + 2\alpha s + \alpha^2$ Hence $2\alpha = 16; \alpha = 8$ $\alpha^2 = 64$

Then the denominator is $(s+8)^2 + 36 = (s+8)^2 + 36$

We know $\mathcal{L}^{-1}$'s $\left\{\dfrac{s+\alpha}{(s+\alpha)^2 + \omega^2}\right\} = e^{-\alpha t} \cos \omega t$ and

$\mathcal{L}^{-1}$'s $\left\{\dfrac{\omega}{(s+\alpha)^2 + \omega^2}\right\} = e^{-\alpha t} \sin \omega t$

and we massage our transform into these forms. We must use all s in the numerator in the cosine transform

$\dfrac{s+20}{(s+8)^2 + 6^2} = \dfrac{s+8+12}{(s+8)^2+6^2} = \dfrac{s+8}{(s+8)^2+6^2} + \dfrac{2 \times 6}{(s+8)^2+6^2}$

Hence $f(t) = e^{-8t}\left[\cos 6t + 2 \sin 6t\right]$

THE S DOMAIN EQUIVALENT CIRCUIT

11-17

Draw the S domain equivalent circuit for the network and initial conditions shown.

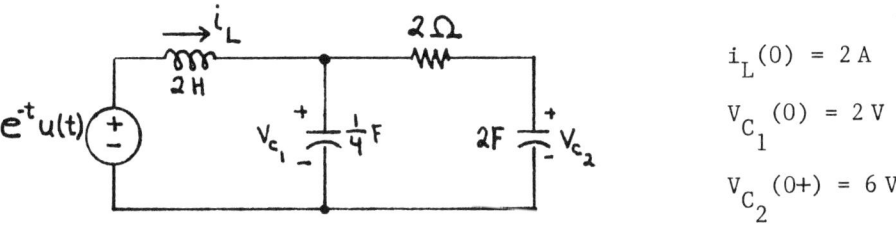

$i_L(0) = 2$ A

$V_{C_1}(0) = 2$ V

$V_{C_2}(0+) = 6$ V

****

There are several solutions, depending upon the initial condition model used for the inductor and the capacitors. We choose (in the time domain) a parallel constant current source for the inductor and series constant voltage sources for the capacitors. In this case, the S domain circuit is:

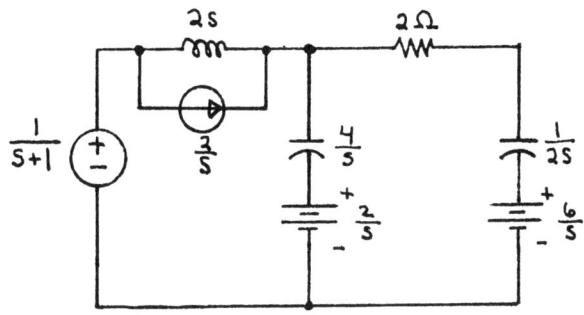

11-18

The following circuit represents the basic form of a ramp-driven TV deflection system. Derive a differential equation in terms of i and solve using Laplace transforms. Sketch i as a function of time.

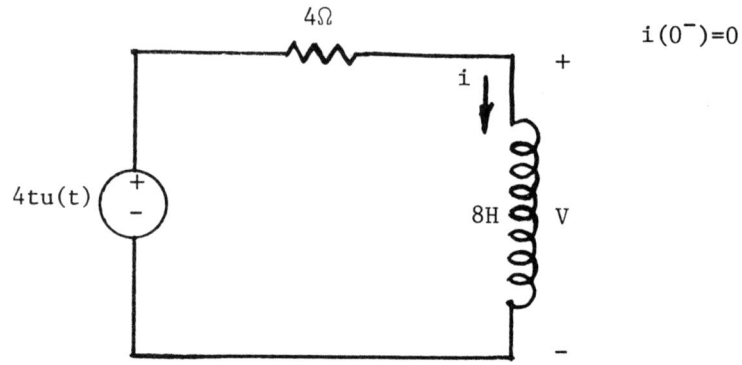

$i(0^-) = 0$

****

By Kirchhoff's Voltage Law,

$$4t\,u(t) = 4i + 8\,di/dt$$

The Laplace Transform of The Equation Is

$$4/s^2 = 4I(s) + 8sI(s)$$

And

$$I(s) = (1/2)/(s^2(s+1/2))$$

This Can Be Expressed In A Partial Fraction Expansion As

$$I(s) = \frac{A}{s} + \frac{B}{s^2} + \frac{C}{s+1/2}$$

The Coefficients Can Be Obtained As Follows:

$$A = d(s^2 I(s))/ds \bigg|_{s=0} = \frac{-1/2}{(s+1/2)^2}\bigg|_{s=0} = \frac{-1/2}{1/4} = -2$$

$$B = s^2 I(s)\bigg|_{s=0} = \frac{1/2}{s+1/2}\bigg|_{s=0} = \frac{1/2}{1/2} = 1$$

$$C = (s+1/2)I(s)\bigg|_{s=-1/2} = \frac{1/2}{s^2}\bigg|_{s=-1/2} = \frac{1/2}{1/4} = 2$$

Then,
$$I(s) = -2/s + 1/s^2 + 2/(s+1/2)$$
And The Inverse Laplace Transform Gives
$$i(t) = -2u(t) + t\,u(t) + 2e^{-t/2}u(t)$$
$$= t\,u(t) - 2(1-e^{-t/2})u(t) \quad A.$$

A Sketch of $i(t)$ Is Given As Follows:

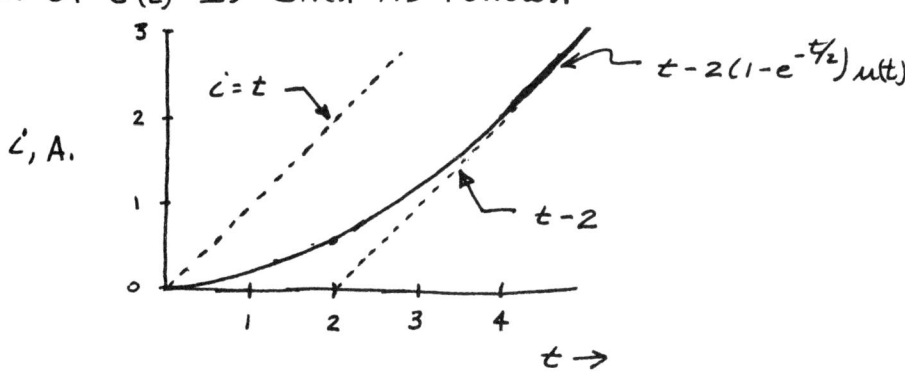

THE COMPLETE RESPONSE OF CIRCUITS

11-19

Using the method of Laplace transform analysis, determine the voltage across resistor R_2 as a function of time.

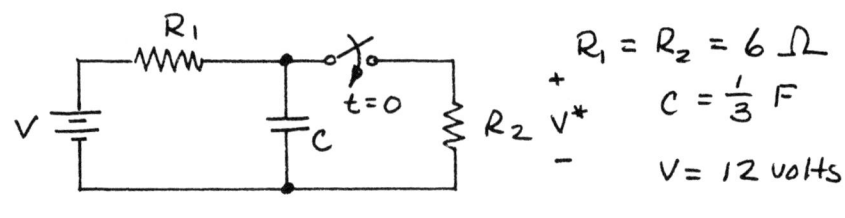

$R_1 = R_2 = 6\,\Omega$
$C = \frac{1}{3}\,F$
$V = 12\text{ volts}$

***

$V_{R_2} = V^*$ and writing a Nodal equation at the top of R_2. For $t \geq 0$, we get

$$\frac{V^* - V}{R_1} + C\frac{dV^*}{dt} + \frac{V^*}{R_2} = 0$$

$$\frac{dV^*}{dt} + \frac{(R_1 + R_2)V^*}{CR_1R_2} = \frac{V}{R_1 C}$$

Substituting values,

$$\frac{dV^*}{dt} + V^* = 6$$

$$sV^*(s) - V^*(0) + V^*(s) = \frac{6}{s}$$

$$V^*(s)(s+1) = \frac{6}{s} + 12 = \frac{6 + 12s}{s}$$

$$V^*(s) = \frac{12s + 6}{s(s+1)} = \frac{A}{s} + \frac{B}{s+1}$$

$$A = V^*(s) \cdot s \bigg|_{s=0} = \frac{12s+6}{s+1} \bigg|_{s=0} = 6$$

$$B = V^*(s) \cdot (s+1) \bigg|_{s=-1} = \frac{12s+6}{s} \bigg|_{s=-1} = 6$$

$$V^*(s) = \frac{6}{s} + \frac{6}{s+1}$$

$$V_{R_2}(t) = V^*(t) = 6\left(1 + e^{-t}\right) u(t)$$

11-20

Use resistors and capacitors in a voltage divider configuration to realize the following transfer functions:

a) $\dfrac{V_2}{V_1} = \dfrac{1}{s+1}$ b) $\dfrac{V_2}{V_1} = \dfrac{s+1}{s+2}$

****

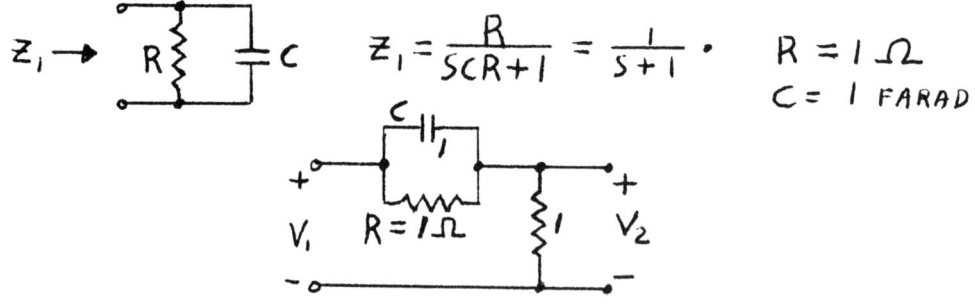

$$\frac{V_2}{V_1} = \frac{Z_2}{Z_1 + Z_2} = \frac{1}{1 + \frac{Z_1}{Z_2}}$$

a) COMPARE $\dfrac{1}{1+\frac{Z_1}{Z_2}}$ WITH $\dfrac{1}{1+S}$. THUS $\dfrac{Z_1}{Z_2} = S$.

CHOOSE $Z_2 = \dfrac{1}{S}$. THEN $Z_1 = 1$.

b) COMPARE $\dfrac{1}{1+\frac{Z_1}{Z_2}}$ WITH $\dfrac{S+1}{S+2} = \dfrac{S+1}{S+1+1} = \dfrac{1}{1+\frac{1}{S+1}}$

THUS $\dfrac{Z_1}{Z_2} = \dfrac{1}{S+1}$. CHOOSE $Z_1 = \dfrac{1}{S+1}$. THEN $Z_2 = 1$.

Z_1 IS IMPEDANCE OF PARALLEL CONNECTION OF RESISTOR AND CAPACITOR.

$Z_1 = \dfrac{R}{SCR+1} = \dfrac{1}{S+1}$. $R = 1\,\Omega$, $C = 1$ FARAD

11-21

The switch has been closed for a long time. It is opened at t = 0. Use Laplace transform methods to find $v_c(t)$ for t > 0.

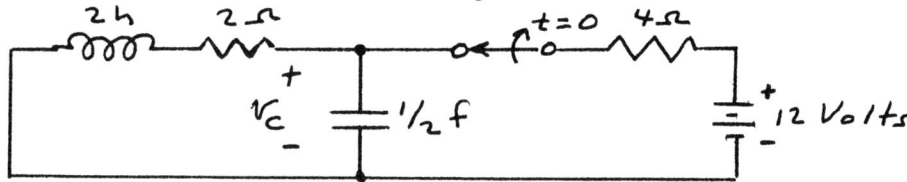

****

For t<0 the capacitor is an open circuit and the inductor is a short circuit

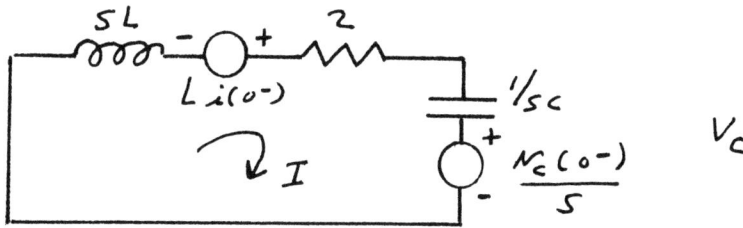

$$i_L(0^-) = -\frac{12}{6} = -2 \text{ Amp}$$

$$v_c(0^-) = 12 \times \frac{2}{2+4} = 4 \text{ Volt}$$

For t>0

$$I = \frac{Li(0^-) - \frac{v_c(0^-)}{s}}{sL + 2 + \frac{1}{sc}} = \frac{LCsi(0^-) - Cv_c(0^-)}{s^2LC + 2sC + 1}$$

$$= \frac{-2s - 2}{s^2 + s + 1}$$

$$V_c = \frac{I}{sc} + \frac{v_c(0^-)}{s} = \frac{-4s - 4}{s(s^2 + s + 1)} + \frac{4}{s} = \frac{4s}{s^2 + s + 1}$$

$$= 4 \left[\frac{s + \frac{1}{2}}{(s + \frac{1}{2})^2 + (\frac{1}{2}\sqrt{3})^2} \right] - \frac{4}{\sqrt{3}} \left[\frac{\frac{1}{2}\sqrt{3}}{(s + \frac{1}{2})^2 + (\frac{1}{2}\sqrt{3})^2} \right]$$

$$\boxed{v_c(t) = 4e^{-t/2} \cos \tfrac{1}{2}\sqrt{3}\, t - \frac{4}{\sqrt{3}} e^{-t/2} \sin \tfrac{1}{2}\sqrt{3}\, t}$$

11-22

Find v(t) for t ≥ 0 if the switch is opened at t = 0.

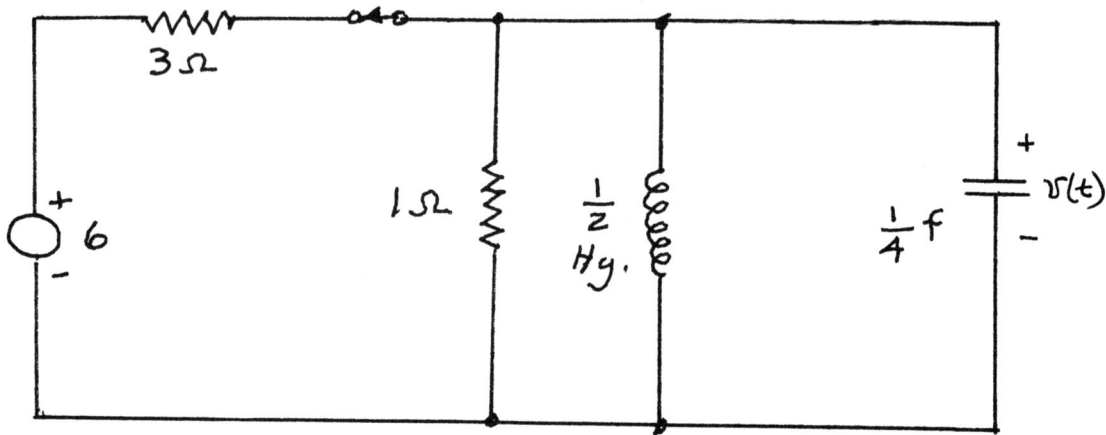

* * * * * * * * * * * * * * * * * * * *

At $t = 0^-$, the current in the inductor is $i_L(0^-) = \frac{6}{3} = 2$ amps

Therefore, $i_L(0^+) = i_L(0^-) = 2$ amps

The circuit for $t \geq 0$ is:

Differential Eqn. is

$$-2 = \frac{v}{R} + C\frac{dv}{dt} + \frac{1}{L}\int_0^t v(t)dt$$

(Note that 2A is the initial inductor current, and it has been transferred to the left hand side of the equality.)

Taking transforms,

$$-\frac{2}{s} = \frac{V(s)}{1} + \frac{1}{4}sV(s) + \frac{2}{s}V(s)$$

$$-\frac{2}{s} = V(s)\left(1 + \frac{s}{4} + \frac{2}{s}\right)$$

Solving, $V(s) = \dfrac{-8}{s^2 + 4s + 8}$

$$V(s) = \frac{-8}{(s+2)^2 + 4}$$

Taking inverse transform,

$$v(t) = -4e^{-2t}\sin 2t, \quad t \geq 0$$

11-23

For the circuit below find $i_2(t)$ for $t \geq 0$ using Laplace transform techniques.

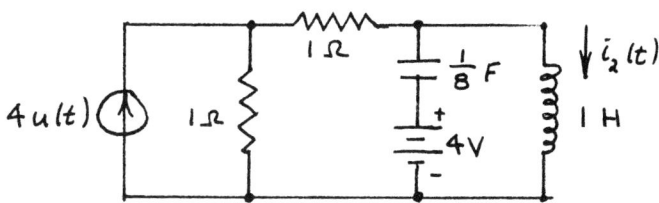

****

For $t < 0$ the circuit looks like

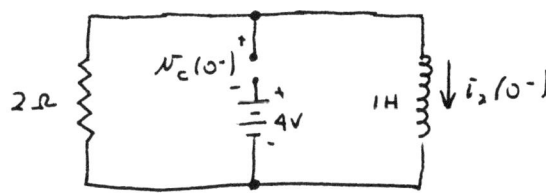

$$v_c(0) = -4 V \qquad i_2(0^-) = 0$$

For $t > 0$ the transformed circuit is:

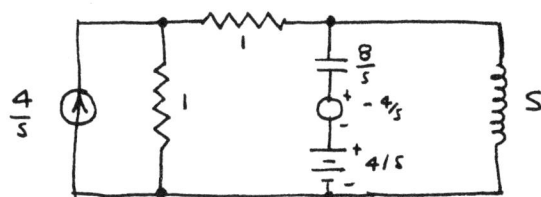

Convert the current source to a voltage source

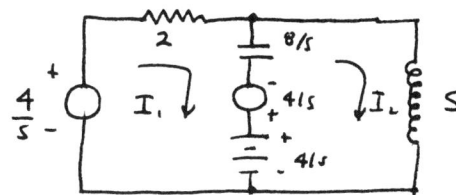

$$\left(2 + \frac{8}{s}\right) I_1 - \frac{8}{s} I_2 = \frac{4}{s} \qquad 2(s+4) I_1 - 8 I_2 = 4$$

$$-\frac{8}{s} I_1 + \left(s + \frac{8}{s}\right) I_2 = 0 \qquad -8 I_1 + (s^2 + 8) I_2 = 0$$

$$(s+4) I_1 - 4 I_2 = 2$$
$$-8 I_1 + (s^2 + 8) I_2 = 0$$

$$I_2(s) = \frac{\begin{vmatrix} s+4 & 2 \\ -8 & 0 \end{vmatrix}}{\begin{vmatrix} s+4 & -4 \\ -8 & s^2+8 \end{vmatrix}} = \frac{16}{s^3 + 4s^2 + 8s + 32 - 32}$$

$$I_2(s) = \frac{16}{s(s^2+4s+8)} = \frac{A}{s} + \frac{Bs+C}{s^2+4s+8}$$

$$As^2 + 4As + 8A + Bs^2 + Cs = 16$$

$$A+B = 0 \qquad 4A+C = 0 \qquad 8A = 16 \qquad A = 2$$

$$B = -2 \qquad C = -8$$

$$I_2(s) = \frac{2}{s} - \frac{2(s+4)}{(s+2)^2+4} = \frac{2}{s} - \frac{2(s+2)}{(s+2)^2+4} - \frac{4}{(s+2)^2+4}$$

$$i_2(t) = 2 - 2e^{-2t}\cos 2t - 2e^{-2t}\sin 2t \qquad \text{for } t \geq 0$$